Conference of Theoretical Physics and Nonlinear Phenomena (CTPNP2018)

AA006214

Journal of Physics: Conference Series Volume 1354

Makassar, Indonesia
2 - 3 July 2018

Editor:

Tasrief Surungan

ISBN: 978-1-7138-0013-2
ISSN: 1742-6588

Printed from e-media with permission by:

Curran Associates, Inc.
57 Morehouse Lane
Red Hook, NY 12571

CURRAN ASSOCIATES INC. proceedings.com

Some format issues inherent in the e-media version may also appear in this print version.

This work is licensed under a Creative Commons Attribution 3.0 International Licence.
Licence details: http://creativecommons.org/licenses/by/3.0/.

No changes have been made to the content of these proceedings. There may be changes to pagination and minor adjustments for aesthetics.

Printed with permission by Curran Associates, Inc. (2020)

For permission requests, please contact the Institute of Physics at the address below.

Institute of Physics
Dirac House, Temple Back
Bristol BS1 6BE UK

Phone: 44 1 17 929 7481
Fax: 44 1 17 920 0979

techtracking@iop.org

Additional copies of this publication are available from:

Curran Associates, Inc.
57 Morehouse Lane
Red Hook, NY 12571 USA
Phone: 845-758-0400
Fax: 845-758-2633
Email: curran@proceedings.com
Web: www.proceedings.com

TABLE OF CONTENTS

PLENARY SESSION

PARALLEL SESSION

Journal of Physics: Conference Series **1354** (2019) 011001 doi:10.1088/1742-6596/1354/1/011001

Conference on Theoretical Physics and Nonlinear Phenomena 2018 (CTPNP2018)

Conference on Theoretical Physics and Nonlinear Phenomena is an annual conference of the Indonesian Theoretical Physicist Group (Grup Fisika Teoretik Indonesia = GFTI). The conference has been an effective gathering place for theorists from Indonesia and neighboring countries to exchange and share ideas. The conference is mainly intended to motivate young physicists to actively do research in the fields. The first meeting (academic gathering) in the form of workshop commenced in 2004, held in Indonesia University. Since 2010, the workshop developed into a conference and named as CTPNP (Conference on Theoretical Physics Nonlinear Phenomena). This annual conference has been held in various major cities of Indonesia, such as Jakarta, Surabaya, Bandung and Jogyakarta. It is for the first time for this year (2018), the conference is held outside Java, here in Makassar, the capital city of South Sulawesi, on the 2nd July 2018, in conjunction with National Physics Seminar on 3rd July 2018, hosted by Physics Department of Hasanuddin University.

As a phyiscs community in this university, we are happy that this important event finally comes to Hasanuddin University. We believe that having this event in Makassar will inspire our physics community in this region. The conference is aimed to strengthen the collaboration and networking among researchers in theoretical physics and nonlinear phenomena, so that members of this group could increase their academic productivity. We hope that this aim can be enlightened by this year CTPNP's main theme: Cooperative phenomena in small and large scale world. Cooperative phenomena are ubiquitous in nature, emphasizing the importance of research collaboration.

As organizer and editor of this proceeding, we would like to take this opportunity to thank all people and institutions who have give significant contribution for realizing this conference, in particular Hasanuddin university for having been a main sponsor. We thank our plenary speakers and all contributive presenters and audiences, coming from different universities and research institutes. Our appreciation is extended to the Head of Physics Department and Dean of Science Faculty for fully support this conference.

Content from this work may be used under the terms of the Creative Commons Attribution 3.0 licence. Any further distribution of this work must maintain attribution to the author(s) and the title of the work, journal citation and DOI.
Published under licence by IOP Publishing Ltd

We also would like to thank the Theoretical Physics division of the Indonesian Physical Society (IPS) who have given the opportunity to the Department of Physics, Hasanuddin University to host this conference.

We wish all participants with a very pleasant and enjoyable time during the conference day and staying in Makassar, which is the gate city for eastern part of Indonesia.

Faithfully,

On behalf of the organizer and Editors

Tasrief Surungan, PhD

Chairman and Chief Editor of CTPNP 2018

Peer review statement

All papers published in this volume of *Journal of Physics: Conference Series* have been peer reviewed through processes administered by the proceedings Editors. Reviews were conducted by expert referees to the professional and scientific standards expected of a proceedings journal published by IOP Publishing.

Content from this work may be used under the terms of the Creative Commons Attribution 3.0 licence. Any further distribution of this work must maintain attribution to the author(s) and the title of the work, journal citation and DOI.
Published under licence by IOP Publishing Ltd

Towards a mean-field kinetic model of electroweak baryogenesis

Andrea Montessori[1], **Tasrief Surungan**[2] **and Sauro Succi** [3⋆]

[1]Department of Engineering, University of Rome "Roma Tre," Via della Vasca Navale 79, 00141 Rome, Italy

[2]Department of Physics, Hasanuddin University, Makassar, South Sulawesi, 90245, Indonesia

[3]Center for life nanoscience at la Sapienza, Istituto Italiano di Tecnologia, viale Regina Elena 295, 00161, Rome, Italy

E-mail: s.succi@iac.cnr.it⋆

Abstract. We explore the dynamic symmetry breaking between the mass of two annihilating species entrapped within a condensing droplet of a third phase-changing species. The symmetry breaking is induced by different diffusivities and/or diffebet source terms acting at the droplet interface. Potential implications for the problem of electroweak baryogenesis are sketched out.

1. Introduction

The problem of baryogenesis, namely the large asymmetry between matter and antimatter observed in the current Universe, is one of the outstanding issues in modern physics and cosmology [1]. A pioneering investigation by Toussaint and Wilczek [2], pointed out that the time asymptotic behaviour of the mutually annihilating species (A and B for convenience) crucially depends on the initial conditions. The rationale is quite intuitive: if the species mix, they react and both disappear according to the irreversible reaction $A + B \rightarrow P$, where P denotes a set of product species which contains neither A nor B. If, on the other hand, by some transport mechanism, they manage to demix or segregate apart, so that the product of their concentrations becomes vanishingly small, then annihilation is quenched, thus spawning a chance for both species to survive much longer than under homogeneous mixing conditions. Besides being of great interest on their own right, the details of such survival may have plenty of applications in chemistry, material science or biology, an example in point being the absorption of drugs within liquid droplets for drug-delivery applications.

In this paper we consider a specific mechanism of segregation associated with the growth of droplets within a phase-changing carrier fluid. By postulating a selective transport of the two species across the droplet interface (membrane), we introduce a symmetry-breaking mechanism which is ultimately responsible for the differential entrapment of the facilitated species (the one with higher transmissivity across the membrane, say A) with respect to the inhibited one (say B). A similar, but more intense, effect is observed by introducing different source terms for the two species. The basic question is: how much mass of species A is entrapped in the growing and moving droplet as a function of time? Once again, this is interesting per-se as a fundamental transport problem in dynamically heterogeneous media,and also for the aforementioned practical purposes. To the best of our knowledge, no detailed account of the hydrodynamic complexity

Content from this work may be used under the terms of the Creative Commons Attribution 3.0 licence. Any further distribution of this work must maintain attribution to the author(s) and the title of the work, journal citation and DOI.
Published under licence by IOP Publishing Ltd

associated with a moving and expanding droplet in the presence of transport and chemical reaction, has ever been discussed. This is precisely the aim of the present work, with prospective focus on electroweak baryogenesis.

2. The transport model

We consider three species A, B and C, where C is a fluid carrier undergoing phase-changes, while A and B are passively transported by C and mutually annihilate through chemical reactions. The three species $k = 1, 2, 3$ obey a continuity equation of the form

$$\partial_t \rho_k + \partial_a(\rho_k u_{k,a}) = r_k + s_k \tag{1}$$

where $a = x, y, z$ runs over spatial dimensions and obeys Einsteins's summation rule. Here, r_k and s_k denote the density change rate due to chemical reactions and generic sources, respectively.

Species A and B share the same mass, which we set to unity by convention, $m_A = m_B = 1$, so that the number and the mass of the species are the same quantity. The two species annihilate at the following rate:

$$R_A = R_B = -\alpha \rho_A \rho_B \tag{2}$$

where α is an adjustable reaction parameter. The C field serves as a carrier for the A and B species and obeys a non-ideal Navier-Stokes equation:

$$\partial_t(\rho_C u_a) + \partial_b P^C_{ab} = s_C \tag{3}$$

where

$$P^C_{ab} = \rho_C u_a u_b + p\delta_{ab} - \sigma_{ab} + \chi \partial_a \rho_C \partial_b \rho_C \tag{4}$$

is the non-ideal momentum flux tensor, including the contributions of inertia, ideal and non-ideal pressure, dissipation and the capillary forces responsible for the first-order phase transition. The right hand side represents an external source of mass.

The species A and B are passively transported by the C-field and diffuse across it with a diffusivity coefficients D_A and D_B respectively. The A and B species experience a selective permeability of the droplet interface, so that an excess of A over B accumulates around the interface and further penetrates within the expanding droplet. The actual amount of mass engulfed within the droplet resulting from such complex transport process is highly sensitive to the chemical details, as well as to the hydrodynamic evolution of the system.

Our main aim is to investigate the complex transport phenomena which result from the dynamic competition between advection, diffusion and reaction processes taking place in the framework of a phase-changing carrier fluid. In our stylized model, microscopic symmetry breaking between species A and B is accounted by introducing different diffusivities and source terms, for the two species A and B, respectively.

2.1. Selective diffusivity

We assume a discontinuous jump between the inner and outer value of the diffusion coefficient of the B species, although smoother dependencies could be easily adjusted. In the LB scheme (see appendix), the diffusivity is controlled by the relaxation frequency ω, hence the jump in diffusivity implies a corresponding change of such frequency across the membrane, The ratio D_{in}/D_{out} of the B-species is the symmetry-breaking parameter to be varied in the source-free simulations.

2.2. Selective sources

We assume that presence of a fourth species, segregated around the droplet interface and giving rise to further production of both species A and B, although at different rates. This effect is modelled by the following source terms:

$$s_A(x, y) = s_0 W(\frac{\vec{r} - \vec{r}_s}{w}) \tag{5}$$

$$s_B(x, y) = \zeta s_0 W(\frac{\vec{r} - \vec{r}_s}{w}) \tag{6}$$

where $W(\cdot)$ is a piece-wise constant centred around the droplet interface of width w, and $0 \leq \zeta \leq 1$ is the symmetry-breaking parameter (SBP).

For standard terrestrial fluids, the hypothetical fourth species could be a solid catalyzer, say a reactive colloid, segregating at the interface, like in bijels. For the case of electroweak baryogenesis (EWBG) we refer to sphalerons, namely hypothetical particles converting baryons to antileptons and antibaryons to leptons, thus changing the baryon number in the process. For a detailed account, see [9].

3. Analytical models

The homogeneous case without space dependence lends itself to some useful analytical considerations which we report in the sequel for both undriven (no sources) and driven case.

3.1. Homogeneous non-driven case

In the homogenous-symmetric scenario $D_A = D_B = 0$, no-phase transitions, no sources and symmetric initial conditions, both species decay according to the nonlinear homogeneous equation:

$$\frac{d\rho_k}{dt} = -\alpha \rho_k^2,$$

whose analytical solution reads as follows:

$$\rho_k(t) = \frac{\rho_{k0}}{1 + \alpha \rho_{k0} t}, \quad k = A, B \tag{7}$$

This yields a τ_k/t decay, where

$$\tau_k \equiv \frac{1}{\alpha \rho_{k0}}, \quad k = A, B \tag{8}$$

is the density-dependent annihilation time-scale.

In the presence of a symmetry-breaking membrane, the two species are expected to develop different values of τ_k, which we refer as to a *dynamic symmetry breaking* due to the effect of the selective interface on the species density. Such dynamic symmetry breaking is expected to occur as soon as the droplet starts to grow, i.e. it starts to nucleate out of its initial seed of radius R_0. Both species A and B begin to be entrapped within the nucleating droplet and their mass within the droplet grows accordingly, as long as the droplet growth rate exceeds their annihilation rate.

As we shall see, such growth is far from monotonic, but characterized instead by large fluctuations, due to the carrier density waves radiating away from the expanding droplet. Such oscillations do not settle down until the droplet condensation has come to an end, i.e at $t \sim \tau_g \equiv R/\dot{R}$, where τ_g defines the condensation time of the droplet, i.e. the time it takes for its mass to reach steady-state.

In the long-term, namely at $t \gg \tau_g$, the densities of the two species are expected to settle to constant values inside and outside the droplet, thus leading to the coexistence of two homogeneous compartments: the droplet and its surrounding environment.

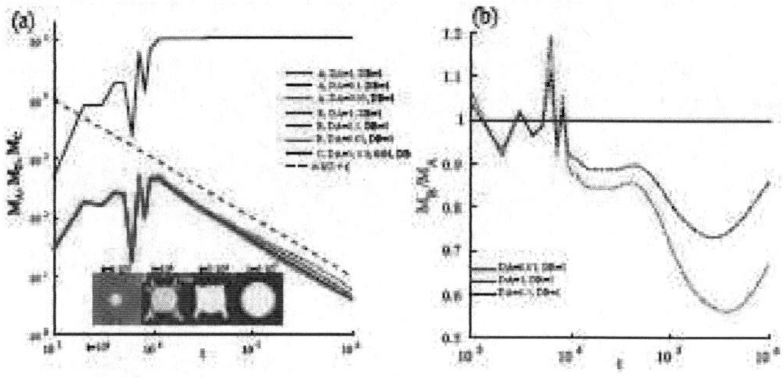

Figure 1. (Color online)Time evolution of the mass of species A, B and C entrapped within the droplet of the carrier species C (Left). The ratio M_B/M_A as a function of time for the three cases $D_{in}/D_{out} = 1, 0.1, 0.01$ (Right). The lower panel at the left reports a time sequence of the density contours of the carrier species C.

Since the droplet is homogeneous, the mass of the entrapped species is expected to follow again a τ_k/t decay, with two different values of τ_A and τ_B, due to the aforementioned dynamic symmetry breaking. In the sequel, we shall put these qualitative considerations on quantitative grounds based on the result of extensive numerical simulations.

3.2. Homogeneous driven case

The equations of the mass evolution within the droplet for the homogeneous driven system read as follows:

$$\dot{M}_A = -\alpha R_{AB} + S_A \tag{9}$$

$$\dot{M}_B = -\alpha R_{AB} + S_B \tag{10}$$

where we have defined the global density overlap

$$R_{AB} = \int \rho_A(x,y)\rho_B(x,y)dxdy,$$

and the global mass inputs per unit time

$$S_k = \int s_k(x,y)dxdy, \; k = A, B.$$

Subtracting the two equations (9) - (10), delivers:

$$\dot{M}_A - \dot{M}_B = S_A - S_B$$

This shows that the mass deficit $M_A - M_B$ grows linearly in time.

By multiplying the first by M_B, the second by M_A and summing them up, we obtain

$$\frac{d}{dt}(M_A M_B) = -\alpha R_{AB}(M_A + M_B) + S_B M_A + S_A M_B$$

The right hand side is made zero by imposing

$$R_{AB} = \frac{1}{\alpha}(S_B\phi_A + S_A\phi_B) \tag{11}$$

where we have defined the mass fractions $\phi_k = M_k/(M_A + M_B)$, $k = A, B$.

Journal of Physics: Conference Series **1354** (2019) 012001 doi:10.1088/1742-6596/1354/1/012001

Next, we write $R_{AB} = \xi \frac{M_A M_B}{V}$, which defines the spatial correlation coefficient, ξ, V being the droplet volume. As a result, eq. (12) yields:

$$M_A M_B = \frac{V\xi}{\alpha} S_{AB} \tag{12}$$

where we have defined $S_{AB} = (S_A \phi_B + S_B \phi_A)$.

Solving explicitly for the time dependence of both masses in the asymptotic regime, we obtain:

$$M_A(t) = s_0 V_{shell} \frac{(1-\zeta)t}{2} (+1 + \sqrt{1 + 4\tau^2/t^2}) \tag{13}$$

$$M_B(t) = s_0 V_{shell} \frac{(1-\zeta)t}{2} (-1 + \sqrt{1 + 4\tau^2/t^2}) \tag{14}$$

where we have defined the decay time scale

$$\tau^2 = \frac{M_A M_B}{(S_A - S_B)^2}$$

which manifestly diverges in the symmetric limit $S_A = S_B$. Note that since the sources are localized around the interface, the volume V_{shell} in the above expression is given by $V_{shell} = Aw$, where A is the area of the droplet surface and w is its thickness.

The upshot of the above analysis is that the majority species grows asymptotically like $\propto t$ and the minority species decreases like $\propto 1/t$. As a result the mass ratio M_B/M_A decays asymptotically in time as:

$$\frac{M_B}{M_A} \sim \frac{\tau^2}{t^2}$$

This is potentially far reaching, since it means that even a minuscule asymmetry in the sources is destined to give rise to a power-law (quadratic) extinction of the minority species. As a mere curiosity, to achieve the current value of the matter/antimatter asymmetry ratio $M_B/M_A \sim 10^{-6}$, with the homogeneous $1/t^2$ decay rate, one should wait a time lapse $10^3\tau$.

Next, let us estimate the decay time as follows. Upon assuming $\phi_B \ll 1$ and $\epsilon \ll 1$, we can write

$$S_{AB} \sim S_B$$

which delivers

$$\tau^2 = \frac{V\xi}{\alpha S_B \epsilon}$$

By writing $S_B = s_0 \zeta V_{shell} \sim s_0 V_{shell}$, we finally estimate

$$\tau^2 = \frac{1}{\alpha s_0} \frac{\xi V}{V_{shell}} \frac{1}{\epsilon^2} \tag{15}$$

This expression invites a transparent physical interpretation. The first term is the product of the source time scale and the annihilation time scale, call it τ_{as}^2. The second term on the right hand side is the thickness parameter of the droplet, essentially the width over its effective radius, where "effective" accounts for the inclusion of the correlation coefficient ξ between the two species. Finally, the third factor is the symmetry breaking source term.

Once τ is assessed, any given antimatter/matter ratio $\mu \equiv M_B/M_A$ is attained on a time scale $\tau_\mu = \tau \mu^{-1/2}$. For the case of EWBG, $\mu = 10^{-6}$ and $\tau_\mu = 10^3\tau$. Of course, these numbers are purely qualitative, but serve nonetheless the purpose of showing that, under a quadratic time decay regime, even a tiny source asymmetry could account for the observed matter/antimatter asymmetry if the sources persists long enough in units of the decay time scale τ.

3.3. Heterogeneous driven case

The previous analysis omits the contribution of diffusion across the droplet membrane, amounting to the mass flux terms $\int D_k \nabla \rho_k \cdot \hat{n} dA$, $k = A, B$. These can be reabsorbed within an effective source term of the form

$$s_k \rightarrow s_k + \frac{D_k}{w} g_k$$

where g_k is the surface-averaged density gradient over the droplet interface. With this renormalisation, the previous analysis carries over with no modifications, except that different numerical values are expected, if the diffusive contributions cannot be neglected. On purely dimensional grounds, diffusive effects are expected to be negligible whenever the diffusive time is longer than the source timescale (strongly-driven regime), i.e:

$$Da_s \equiv \frac{w^2}{D\tau_s} \gg 1$$

where Da_s is the so-called Damkohler number associated with the source term. An analogue Damkohler number applies to annihilation, the condition

$$Da_a \equiv \frac{w^2}{D\tau_a} \gg 1$$

denoting the situation in which the species experience substantial annihilation while traversing the droplet interface (fast-annihilation regime).

4. Numerical set-up

In the sequel we introduce the simulation set-up and present numerical results for the selective source scenario.

4.1. Initial and boundary conditions

Both species are initialised at the same constant density value throughout the computational domain:

$$\rho_A = \rho_B = \rho_0 = 1 \tag{16}$$

The initial density of the carrier field is defined as follows:

$$\rho_C = \rho_{in}(1 \pm \eta), \text{ in a ball of radius } R_0 \tag{17}$$

$$\rho_C = \rho_{out} \text{ outside the ball} \tag{18}$$

where η is a zero-mean random perturbation with rms $\delta\rho/\rho = 0.01$. We take $\rho_0 = 1$, $\rho_{in} = 0.2$, $\rho_{out} = 0.10$ and $R_0 = 10$.

For the phase transition, we choose $T/T_c = 0.04/0.047$, corresponding to a coexistence liquid/vapour density ratio of about $0.27/0.027 = 10$. The values of ρ_{out} and ρ_{in} determine the duration of the growth stage, i.e. the time it takes for the droplet to attain the coexistence values of the liquid (l) and vapour (v) phases.

By mass conservation:

$$\rho_{in} V_{in} + \rho_{out} V_{out} = \rho_l V_l + \rho_v V_v = M \tag{19}$$

where $V_{in} + V_{out} = V_l + V_v = V = L^2$ is the total volume of the system. Clearly, the volume of the droplet grows at increasing the total mass in the system. More specifically, the final value of the volume fraction, i.e. the ratio of the volume of the liquid droplet to the total volume, is given by:

$$\lambda \equiv \frac{V_l}{V} = \frac{\rho_m - \rho_v}{\rho_l - \rho_v} \tag{20}$$

where $\rho_m = M/V$ is the average mass density. The diameter of the liquid droplet is thus given by:

$$D = L\sqrt{\frac{4}{\pi}\lambda} \tag{21}$$

Hence, the maximum droplet diameter, $D = L$, is attained at the a volume fraction $\lambda_{max} = \sqrt{\pi/4} \sim 0.87$.

All species are taken initially at rest, namely:

$$\vec{u}_k = 0, \; k = A, B, C \tag{22}$$

and full periodicity is assumed across the four boundaries of the simulation box.

Collisional time-scales are taken as $1/\omega_A = 1/\omega_B = 1/\omega_C = 1$, fixing the fastest timescale in action. This corresponds to an outer diffusivity and carrier viscosity, $D_A = D_B = \nu_C = 1/6$ in lattice units (see Appendix). The annihilation rate is taken as $\alpha = 0.1$, corresponding to an annihilation timescale $\tau_a = 10$ at unit density $\rho_k = 1$.

5. Numerical results

We consider a two-dimensional square with 1024 grid-points per side and run over a timespan of 10^6 time-steps, thus covering three decades in space and six in time, as it is appropriate for diffusive phenomena. The simulations have been run using a three-component lattice Boltzmann scheme (see Appendix).

To fix ideas for prospective applications to EWBG, let us inspect the physical time and length scales of the simulations. The time span goes from the onset of EWBG, $t_{EWBG} \sim 10^{-11}$ seconds, to the time of the QCD transition, $t_{QCD} \sim 10^{-5}$ seconds. With one million timesteps, this fixes the lattice timestep to $\Delta t = t_{EWBH} = 10$ ps. The corresponding lattice spacing is $\Delta x = c\Delta t = 3\ 10^{-3}$ meters, which means that we deal with a computational Universe of side $L = 3$ meters, and a Higgs droplet inflating from about 3 mm to 3 meters in diameter.

5.1. *Source-free simulations*

In Figure 1 we present the evolution of the mass of the three species A, B, C for the source free case and different combinations of the diffusion coefficients, namely, outside the droplet $D_A = D_B = 1/6$, while inside $D_B = 1/6, 1/60, 1/600$, so as to break the A-B symmetry [10].

As one can see, after an initial transient, showing large fluctuations (not shown) , both species A and B settle close to the expected τ_k/t decay and with $\tau_A \neq \tau_B$, due to the dynamic symmetry breaking occurring in the transient. By and large, however, the mass ratio M_B/M_A remains $O(1)$, showing that the diffusive symmetry breaking does not appear to be relevant to the EWBG scenario.

5.2. *Source-driven simulations*

Next, we investigate the effects of selective source terms for the species A and B. by changing the asymmetry source coefficient in the range $0.4 \leq \zeta \leq 1$. The main parameters are set as follows: $D_A = 1/6, D_B = 1/6$ and $s_0 = 10^{-3}$. With these parameters, given a radius $R \sim 400$ and a width ~ 5, we compute a a decay relaxation time $\tau^2 \sim 4\ 10^5/\epsilon^2$, with $\xi = 1$.

In Figure 2, we present the time evolution of the mass A, B and C within the droplet for the case $\zeta = 0.99$. As clearly shown in the figure, the source term leads to a secular growth of the

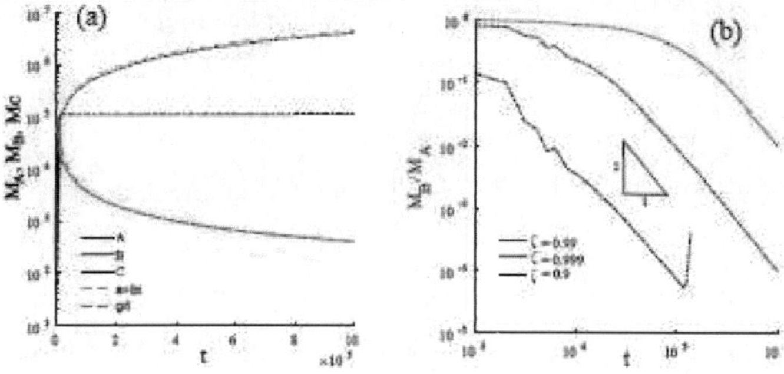

Figure 2. Time evolution of the masses of species A,B and C inside the droplet for the case $\zeta = 0.99$. Panel (b) reports the corresponding mass ratio M_B/M_A as a function of time for the cases $\zeta = 0.999, 0.99, 0.9$ (top to bottom). The whisker in the lower curve is due to a crash of the simulation, due to the large separation between the two species, but the t^{-2} scaling is nonetheless apparent.

A species, $M_A(t) \propto t$, versus a homogeneous $1/t$ decay of the B species. As a result, the mass ratio M_B/M_A goes indeed asymptotically to zero like t^{-2}.

The time asymptotic behaviour sets in long before the density configuration in space reaches its mechanical equilibrium, the chief condition being that the mass of the droplet be stationary in time, regardless of its shape. Since the majority species follows a linear trend $M_A(t) \propto t$, while the minority one obeys a reciprocal trend $M_B(t) \propto 1/t$, their product remains basically asymptotically constant in time, which is indeed confirmed by the numerical simulations (not shown), supporting the assumption made in the analytical model. In particular, with $\epsilon = 0.001, 0.01, 0.1$, we obtain $M_B/M_A \sim 10^{-2}, 10^{-4}, 10^{-6}$ after 10^6 steps, respectively, which is consistent with a correlation coefficient $\xi \sim 0.02$.

6. Relevance to electroweak baryogenesis

It is of interest to speculate whether the present transport model can be of any interest in the context of electro-weak baryogenesis [6]. To this purpose, let us remind that in our model baryogenesis implies the identification *A=matter, B=antimatter and C=Higgs fluid.* It is therefore of interest to assess the plausibility of present model towards the basic requirements for baryogenesis, as as formulated by Sakharov, namely [7]:

i) The existence of an explicit baryon-symmetry breaking mechanism,
ii) Violation of C and CP invariance,
iii) Thermodynamic non-equilibrium

As to i), the baryon symmetry breaking is expressed by the non-unit diffusion jump factor J_{AB} across the membrane, or an explicit symmetry breaking at the level of source terms, i.e $\zeta \neq 1$. The microscopic origin of such symmetry breaking is not relevant to the present mean-field model. Item ii) states that the system must be invariant upon a reflection in space, say from x to $-x$ across the interface, and charge conjugation. Our model is electrically neutral, hence item i) is basically a requirement that the density of A at location x be different fom the density of B at the mirror location $-x$, namely $\rho_A(x) \neq \rho_B(-x)$. This is certainly true once a non-unit diffusivity jump or source asymmetry factor is in action.

Finally, thermodynamic non-equilibrium implies that species A and B must depart from their local thermodynamic equilibrium, which is certainly true in the presence of density gradients

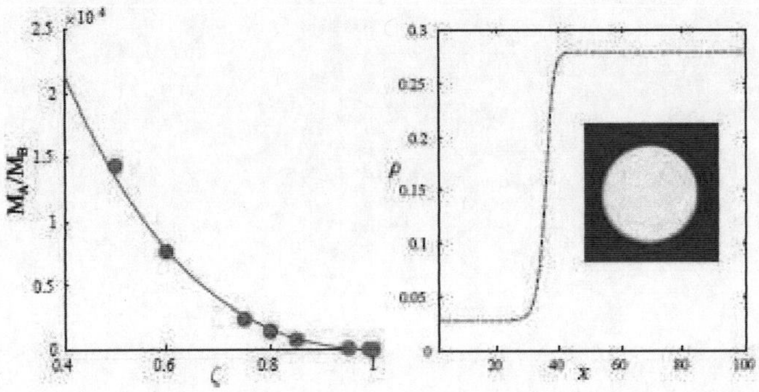

Figure 3. The ratio M_A/M_B at $t = 12000$, as a function of ζ at the time when the droplet attains its equilibrium mass (left). The solid line is the fit $2\ 10^5 e^{-2.3x}\ (1-x)^2$. On the right side, we show a typical density profile of the carrier species across the interface.

across the interface. Thus, at least at a qualitative level, the present model appears compatibel with the basic requirements for baryogenesis.

To proceed towards a quantitative analysis, several aspects need to be inspected in detail. First, in the EWBG scenario the Higgs droplet expands much faster than the Universe, until it fills it up entirely, whence its alleged pervasiveness at the current day. In our model, the droplet stops growing once mass equilibrium is attained, typically for density ratios around 10 between the liquid and vapour phases. In addition, our computational Universe is static, as opposed to an expanding Universe. However, both limitations could be significantly mitigated, if needed.

The droplet growth rate in our simulations is $\dot{R} \sim 0.04$ in light speed units, which is about ten times smaller than the credited wall speed of the true Higgs droplets, estimated at $c/2$. Given that no fine-tuning effort has been spent in customising the simulations to the EBWG scenario, the above figures do not appear completely off target. The thickness parameter $w/R \sim 0.01$ also appears reasonable.

Next, let us inspect the values of the matter/antimatter ratio, in our case the ratio M_A/M_B at the time when the droplet reaches its equilibrium mass. In Fig. 3 we report the ratio M_A/M_B at the end of the droplet growth, as a function of the symmetry breaking parameter ζ. We note that with $\zeta \sim 0.5$, ratios around 10^4 are obtained, which extrapolate to 10^5 in the limit $\zeta \to 0$. These values are two (one) orders of magnitude above the current value of the matter/antimatter ratio in the Universe, which is estimated at about 10^6. Although not visible on the scale of the plot, $\zeta = 0.95$ yields a mass ratio around 80, namely two orders of magnitude, in the face of a tiny five percent source asymmetry.

In addition, as shown earlier on, such ratio grows quadratically in time, hence it is very sensitive to the epoch at which sources fade away. As discussed previously, under quadratic decay, the time at which a ratio 10^{-6} is obtained is $10^3\tau$. By choosing $\tau \sim t_{EWBG}$, this is basically the geometric mean between the time of EWBG and QCD time.

Summarizing, it appears reasonable to speculate that, with proper fine-tuning, the present model could prove useful for computational explorations of the strong non-equilibrium spacetime dynamics of EWBG scenarios.

7. Conclusions and outlook

Summarizing, we have analysed the transport of mutually annihilating species within the flow field of a passive carrier experiencing a first-order dynamic phase transition. In particular, we analysed the symmetry-breaking effects on the mass engulfed by the growing droplet as induced by an explicit symmetry breaking of the diffusive and/or source terms. The source-driven

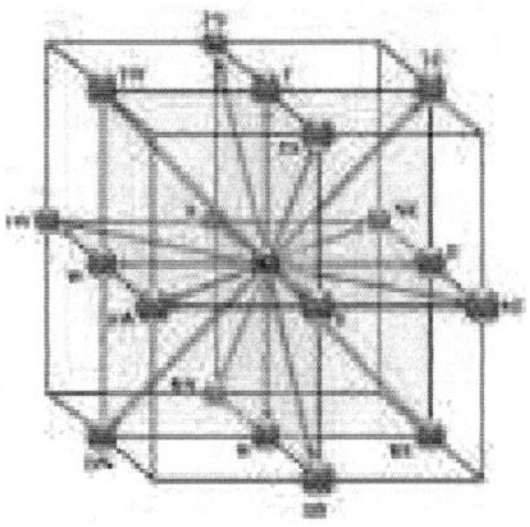

Figure 4. The nineteen discrete velocity lattice in three spatial dimensions (D3Q19).

scenario appears very effective, since any nonzero asymmetry between the two sources turns the $1/t$ decay of the majority species into a secular linear growth. As a result, in the long term, the ratio between minority and majority species decays like $1/t^2$. With suitable adaptations, particularly a more specific modelling of the source terms, the present scheme may permit to gain insisghts into the strongly non-equilibrium spacetime dynamics of electroweak baryogenesis at a very affordable computational cost.

Acknowledgements
The research leading to these results was funded by the European Research Council under the European Union Horizon 2020 Framework Programme (No. FP/2014-2020)/ERC Grant Agreement No. 739964 (COPMAT). One of the authors (SS) acknowledges illuminating discussions with Gian Francesco Giudice, Gino Isidori and David Spergel. He also wishes to thank Fabiola Gianotti for arranging a memorable visit at CERN, during which part of this work was discussed.

Appendix: The Lattice Boltzmann formulation

The transport equations described above are solved by means of a three-species lattice Boltzmann (LB) scheme. The main reason for using LB is its ability of dealing with dynamic phase-transitions in a much handier way than solving the Navier-Stokes equations of non-ideal fluids.

The LB equation takes the following form [4]:

$$f_i^k(\vec{r}+\vec{c}_i\Delta t, t+\Delta t) - f_i^k(\vec{r},t) = -\omega_k \Delta t\,(f_i^k - f_i^{k,eq}) + F_i^k \Delta t,\; k = A, B, C \tag{23}$$

where $f_i(\vec{r};t)$ represents the probability to find a representative particle of species k at the lattice position $\vec{r}$ and time t with the discrete velocity $\vec{c}_i$. The index i runs over the discrete speeds, $i = 0, 18$ for the present nineteen-velocity three-dimensional lattice.

The local equilibria (a truncated version of Maxwell-Boltzmann distribution) encode the mass-momentum conservation laws. At the moment they are purely classical, but readily extend to quantum statistics. In detail:

$$f_i^{A,eq} = w_i \rho_A (1 + u_i) \tag{24}$$

$$f_i^{B,eq} = w_i \rho_B (1 + u_i) \tag{25}$$

$$f_i^{C,eq} = w_i \rho_C (1 + u_i + q_i) \tag{26}$$

$$\tag{27}$$

where $u_i = \vec{u} \cdot \vec{c}_i / c_s^2$ and $q_i = u_i^2 - u^2$, u being the magnitude of the net flow of the carrier fluid, namely

$$\rho \vec{u} = \sum_i f_i^C \vec{c}_i \tag{28}$$

with

$$\rho = \sum_i f_i^C \tag{29}$$

the carrier density. Finally, w_i is the standard set of weights normalised to unity and $c_s^2 = \sum_i w_i c_i^2/d$ is the sound speed in d spatial dimensions. In the present lattice $c_s^2 = 1/3$ (Note that the speed of light is $c = 1$ in lattice units).

The transport properties are controlled by the relaxation rate, according to the standard LB relations, namely:

$$D_A = c_s^2(1/\omega_A - \Delta t/2) \tag{30}$$
$$D_B = c_s^2(1/\omega_B - \Delta t/2) \tag{31}$$
$$\nu_C = c_s^2(1/\omega_C - \Delta t/2) \tag{32}$$

Note that A and B equilibria conserve only mass, hence they support mass diffusion, whereas carrier equilibria conserve momentum as well because the local equilibria contain the self-consistent carrier current, see eq. (28). Consequently the carrier relaxation rate controls momentum diffusivity, also known as kinematic viscosity.

For species A and B, the forcing terms are set to zero $F_i^A = F_i^B = 0$, so that they obey an ideal equation of state

$$p_{A,B} = \rho_{A,B} c_s^2.$$

Given that $c_s^2 = 1/3$ in lattice units, where $c = \Delta x/\Delta t = 1$ is the light speed, this is the equation of state of ultra-relativistic matter, i.e. $mc^2/kT \ll 1$, which is reasonable for EWBG, which operates in the regime $kT \sim 100$ GeV.

The carrier fluid, however, is subject to self-consistent force resulting from potential energy interactions, according to the standard LB pseudo-potential formulation [5]. Consequently, it obeys a non-ideal equation of state of the form (Carnahan-Starling) [11]

$$p_C = \rho_C k_B T \, \frac{1 + \rho + \rho^2 + \rho^3}{(1-\rho)^3} - 0.5\rho^2 \tag{33}$$

corresponding to a critical temperature $T_c = 0.047$ and density $\rho_c = 0.066$ in lattice units.

References

[1] M. Tanabashi et al. (Particle Data Group), Phys. Rev. D 98, 030001 (2018).
[2] D. Toussaint and F. Wilczek, Particle-antiparticle annihilation in diffusive motion, J. Chem. Phys, 78, 2642, (1983).
[3] Electroweak baryogenesis, D.E. Morrissey and M. J. Ramsey-Musolf, arXiv:1206.2942v1 [hep-ph] 13 Jun 2012.
[4] S. Succi, The lattice Boltzmann equation for complex states of flowing matter, Oxford U.P., 2018
[5] X. Shan and H. Chen, *Phys. Rev.* E **47**(3), 1815, (1993).
[6] L.S. Kisslinger, Astrophysics and the evolution of the Universe, World Scientific, (2014).
[7] A. D. Sakharov, Violation of CP Invariance, C Asymmetry, and Baryon Asymmetry of the Universe, Pisma Zh. Eksp. Teor. Fiz. 5, 32 (1967) [JETP Lett. 5, 24 (1967 SOPUA,34,392- 393.1991 UFNAA,161,61-64.1991)
[8] AD Linde, Phase transitions in gauge theories and cosmology, Reports on Progress in Physics 42 (3), 389958 (1979).
[9] M. Trodden, Electroweak baryogenesis, Review of Modern Physics, 71, 1463, (1999).
[10] S. Succi and A. Montessori, *submitted*, (2019)
[11] N. F. Carnahan and K. E. Starling, J. Chem. Phys. **51**, 635 (1969).

Journal of Physics: Conference Series **1354** (2019) 012002 doi:10.1088/1742-6596/1354/1/012002

A Brief Review on Canonical Loop Quantum Gravity: The Kinematical Part

Freddy Permana Zen[1,2], **Seramika Ariwahjoedi**[1]

[1]Theoretical Physics Laboratory, THEPI Division, Institut Teknologi Bandung, Jl. Ganesha 10 Bandung 40132, West Java, Indonesia.

[2]Indonesia Center for Theoretical and Mathematical Physics (ICTMP), Indonesia.

E-mail: `fpzen@itb.ac.id`
`ceramyca@gmail.com`

Abstract. In this article, we briefly review the kinematical part of canonical loop quantum gravity. This article starts with tetradic formulation of gravity both in the covariant approach and canonical approach. The next step is to introduce Ashtekar new variables and to apply the Dirac canonical quantization procedure of gravity. By the holonomy representation, one obtains the loop representation of quantum gravity.

1. Introduction

The search for quantum gravity as a quantization of General Relativity has been carried for a long time and had not been completed recently. The research has evolved in various ways, ranging from the perturbative to the non-perturbative theory. The perturbative quantization meets several obstacles, one of them being the non-renormalizable property of the gravitational field. In the other hand, the non-perturbative theories provide possible ways to quantize gravity; they could be broadly categorized into two main branches: string and non-string approach. One of the candidates of non-string approach of quantum gravity is loop quantum gravity (LQG), which is based rigorously on Dirac quantization procedure.

In this article, we review the basic of canonical loop quantum gravity. Due to time and length constraint, we discuss only on the kinematical part. Section II consists of tetradic formulation of gravity as an attempt to treat gravity as a gauge theory. It contains three subsections, reviewing on the covariant approach, canonical approach, and Ashtekar new variables. In this section, we briefly review the Hamiltonian formulation of general relativity. The third section consists the Dirac canonical quantization procedure of gravity. It contains two subsections, which are the quantization via connection and holonomy representation. The previous are problematic, while the latter leads to loop quantum gravity. Section 4 consist the main subject of this article, namely a review of the kinematical part of loop quantum gravity. It consist four subsections which respectively review the cylindrical functions and the Hilbert space of quantum gravity, the Gauss constraint and the kinematical Hilbert space, the graph and spin-network states, and the geometric operators on quanta of space.

The review on this article is mainly based on reviews of the kinematical part of quantum gravity in [1, 2, 3, 4], and can be viewed as a summary of these articles.

Content from this work may be used under the terms of the Creative Commons Attribution 3.0 licence. Any further distribution of this work must maintain attribution to the author(s) and the title of the work, journal citation and DOI.
Published under licence by IOP Publishing Ltd

Journal of Physics: Conference Series **1354** (2019) 012002 doi:10.1088/1742-6596/1354/1/012002

2. First Order Formulation Of Gravity

2.1. Covariant Approach

The original and standard metric formulation of General Relativity is usually known as the second order formulation. One could start from the Einstein-Hilbert Lagrangian density $\mathcal{L}_{EH}[g] = R(g)$, such that Einstein-Hilbert action of gravity is:

$$S_{EH}[g] = \Delta\, R(g)\ {}^4\text{vol} = \Delta \star R(g),$$

with R is the Ricci scalar, and metric g as the dynamical variable. Minimizing the variation of action δS with respect to a variation on the metric δg, gives the Euler-Lagrange equation which is equivalent to the vacuum Einstein Field Equation:

$$G_{\mu\nu} = R_{\mu\nu} - \frac{1}{2} R g_{\mu\nu} = 0.$$

It is already mentioned in the begining of this article that gravity could be described as a gauge field, and it would be convenient to have General Relativity rewritten in a form closer to gauge theory. In the latter, a 'physical' field is defined as a section of a bundle. In the case of GR, the bundle is a tangent bundle. The first attempt on this approach is the Einstein-Cartan formulation, or non-coordinate base version of GR, where one has bundle $\mathcal{E} \simeq \mathcal{M} \times \mathcal{F}$, equipped with a (spin) connection $\omega \in \mathfrak{sl}(2, \mathfrak{c}) \times \bigwedge^1 (\mathcal{M})$ on spacetime $\mathcal{M}$. One defines a local trivialization $\mathbf{e}$ to map the vector bundle to tangent bundle $T\mathcal{M}$. Using this map, one could send all properties on $T\mathcal{M}$ to $\mathcal{E}$ and vice-versa, for example, the frame field or tetrad $e_I = \mathbf{e}(\xi_I) = e^{\mu}_{I} \partial_\mu$, satisfying the orthogonality condition $\langle e_I, e_J \rangle = e^{\mu}_{I} e^{\nu}_{J} q_{\mu\nu} = \eta_{IJ}$. The Riemann tensor on $\mathcal{M}$ is related to the curvature 2-form of E by $R = \mathbf{e}(F)$, so that the Einstein-Hilbert action can be written as:

$$S_{EH}[e] = \Delta\, e^{\mu}_{I} e^{\nu J} F^{I}_{\mu\nu J}\ {}^4\text{vol} = \frac{1}{2} \varepsilon_{IJKL} \Delta\, e^I \wedge e^J \wedge F^{KL}, \tag{1}$$

where now the tetrad e_I is the dynamical variable. In fact, it is more convenient to consider both of the tetrad and spin connection ω as the dynamical variables, such that the action becomes:

$$S_P[e, \omega] = \Delta\, \mathcal{L}_P[e, \omega] = \frac{1}{2} \varepsilon_{IJKL} \Delta\, e^I \wedge e^J \wedge F^{KL}(\omega),$$

known as the Palatini action [5]. The connection ω enters from the curvature 2-form $F = d_D \omega$. This is known as the first order formulation of gravity.

Minimizing the variation of δS with respect to δe and $\delta \omega$ gives the following Euler-Lagrange equations:

$$\varepsilon_{IJKL} e^J \wedge d_D e^J = 0, \qquad \varepsilon_{IJKL} e^I \wedge F^{KL}(\omega) = 0. \tag{2}$$

If $e^I \neq 0$, the first equation gives the torsionless condition $d_D e^I = 0$, where the solution gives the Levi-Civita connection. The second is the equivalent of the vacuum Einstein Field Equation. Both relations provide the standard dynamics of GR.

2.2. Canonical Approach

As a first step towards the quantization of a system, one needs to obtain the canonical Hamiltonian. For a covariant system, the first step is to apply the 3+1 ADM formulation. The original 3+1 decomposition is done in the metric formulation, this can be seen in [6]. In this article we will only review the tetradic ADM formulation. Let $\mathcal{M}$ be a globally hyperbolic manifold such that local diffeomorphism $\mathcal{M} \to \Sigma \times \mathbb{R}$ exists for each point on $\mathcal{M}$. Let us define $\vec{n}$ as the vector field normal to Σ, the 'time' direction for spatial space Σ. The split $\mathcal{M} \to \Sigma \times \mathbb{R}$

induces the splitting on the fibre $\mathcal{F}$, such that the 3-connection on Σ is ${}^3\omega \in \mathfrak{su}(2) \times \bigwedge^1(\Sigma)$. The derivative of connection and its corresponding momentum conjugate are, respectively :

$$\frac{d^3\omega}{d\tau} = {}^3\dot{\omega}, \qquad \frac{\delta \mathcal{L}_{EH}}{\delta^3\dot{\omega}} = E,$$

with $E = \sqrt{\det q}\ {}^3e$ is the densitized triads and q is the 3-metric of Σ. Using the time gauge as a gauge fixing, and applying the Hodge star operator on the internal indices of the 3-connections and triads, one obtains:

$$\frac{\delta \mathcal{L}}{\delta^3\dot{\omega}^i_a} = E^a_i = e = \frac{1}{2}\varepsilon_{ijk}\varepsilon^{abc}e^j_b e^k_c.$$

The 3-connection and densitized triads are canonically conjugate to each other, such that they satisfy the symplectic structure of the phase space variables from Poisson Brackets:

$$\left\{{}^3\omega^i_a(x), E^b_j(x')\right\} = \delta^i_j\delta^b_a\delta^3(x-x').$$

The corresponding Hamiltonian density could be obtained from the Legendre transformation:

$$\mathcal{H} = e^\mu_I e^{\nu J}F^I_{\mu\nu J} - E^{a3}_i\dot{\omega}^i_a = \omega^i_0\mathcal{G}_i + N\mathcal{C} + N^a\mathcal{C}_a \approx 0.$$

Gravity is a totally constrained system; the Hamiltonian is a summation of constraints density, and therefore $\mathcal{H} \approx 0$. Writing the constraint explicitely [8]:

$$\mathcal{G}_i = \varepsilon_{ij}{}^k K^j_a E^a_k \approx 0, \tag{3}$$
$$\mathcal{C}_a = D_{[b}\left(K^i_{\ a]}E^a_i\right) \approx 0. \tag{4}$$
$$\mathcal{C} = \frac{1}{\sqrt{q}}\left(E^b_iE^a_j - E^a_iE^b_j\right)K^i_aK^j_b - \sqrt{q}^3R \approx 0. \tag{5}$$

$\mathcal{H}$ is the Hamiltonian density; to obtain the Hamiltonian of the system, the corresponding constraint are smeared on the spatial hypersurface:

$$G(\vec{\omega}) = \Delta_\Sigma\, \omega^i_0\mathcal{G}_i q^{12}d^3x, \tag{6}$$
$$C(N) = \Delta_\Sigma\, N\mathcal{C}q^{12}d^3x, \tag{7}$$
$$C\left(\vec{N}\right) = \Delta_\Sigma\, N^a\mathcal{C}_a q^{12}d^3x. \tag{8}$$

$G(\vec{\omega})$, $C\left(\vec{N}\right)$, and $C(N)$ are, respectively, the Gauss, diffeomorphism, and Hamiltonian constraints.

For constrained system, it is important to check that the constraints are consistent to each other. This lead Dirac and Bergmann to propose an algorithm to check the consistency of the constraints [7]. Constraints on a Hamiltonian system are classified into the following conditions: (1) primary constraint, if a constraint is independent from the dynamical equation, otherwise it is called secondary; (2) first class, if a constraint commutes with all other existing constraints on the constraint hypersurface, otherwise it is called second class. First class constraints generates gauge transformation [7, 8].

The constraints in GR (6)-(8) are primary, but second class. To do a quantization on each foliation of the constraint hypersurface, it is more convenient to have a set of first class constraint, which is done in the next subsection.

2.3. Ashtekar New Variables

The procedure proposed by Ashtekar, is an attempt to make the constraints (6)-(8) first class [9]. As another advantage to this, the constraints can be rewritten in nice polynomial forms, which make them easier to quantize.

2.3.1. Holst Term and Holst Action. A Lagrangian of a system is not unique, in the sense that it can be modified as long as the Euler-Lagrange equation is invariant. Using this fact, one could add an additional term labeled as Holst term in the Palatini action [10]:

$$S_H\left[e,\omega\right] = \Delta\,\mathcal{L}_H = \Delta\,\frac{1}{2}\varepsilon_{IJKL}e^I \wedge e^J \wedge F^{KL}\left(\omega\right) + \underbrace{\frac{1}{\gamma}\delta_{IJKL}e^I \wedge e^J \wedge F^{KL}\left(\omega\right)}_{\text{Holst term}}, \tag{9}$$

with $\delta_{IJKL} = \delta_{I[K}\delta_{L]J}$. The factor γ is a constant known as the Barbero-Immirzi parameter, and the action is called as Holst action [10].

Minimizing the variation of δS for any δe and $\delta\omega$ gives the following Euler-Lagrange equations:

$$\left(\frac{1}{2}\varepsilon_{IJKL} + \frac{1}{\gamma}\delta_{IJKL}\right) e^J \wedge F^{KL}\left(\omega\right) = 0, \qquad \left(\frac{1}{2}\varepsilon_{IJKL} + \frac{1}{\gamma}\delta_{IJKL}\right) e^I \wedge d_D e^J = 0.$$

The dynamical equation derived from the Holst action are still equivalent to EFE (2) as long as the tetrads are non-degenerate, $e^I \neq 0$.

In the original version, one starts with a complexification of $\mathfrak{sl}(2,\mathfrak{c})$ by setting $\gamma = i$, thus gauge group of general relativity is the complexified Lorentz group $SO_{\mathbb{C}}(3,1)$. With $SO_{\mathbb{C}}(3,1)$, one has two distinct $SO(3,1)$ elements (which consist the real $SO(3,1)$ group, its complex conjugate part, and their dual groups). But due to the reality issue, it is more convenient to use real variables, that is, restricting γ to be real.

2.3.2. Ashtekar Variables. As already been discussed in the previous subsection, the triads ${}^3\omega$ is an element of $\mathfrak{su}(2) \times \bigwedge^1(\Sigma)$. Let us construct another variables which is also an element of $\mathfrak{su}(2)$, with the following transformation:

$$A = {}^3\omega \pm \gamma \star {}^3\omega \in \mathfrak{su}(2) \times \bigwedge^1(\Sigma) \simeq \mathcal{A}, \tag{10}$$

$\star{}^3\omega$ is the Hodge-dual part (in the internal/fibre space) of ${}^3\omega$. A is the half of the Ashtekar 'new' variables [9]. Taking the derivative of connection, and using the Holst Lagrangian density (9), the corresponding momentum conjugate is exactly the densitized triads of the previous case:

$$\frac{dA}{d\tau} = \dot{A}, \qquad \frac{\delta\mathcal{L}_H}{\delta^3\dot{A}} = E.$$

Using the time gauge and applying the Hodge star operator on the 3-connections and triads, one obtains:

$$\frac{\delta\mathcal{L}_H}{\delta\dot{A}^i_a} = E^a_i = \frac{1}{2}\varepsilon_{ijk}\varepsilon^{abc}e^j_b e^k_c.$$

The symplectic structure is satisfied by $\tilde{E}^b_j = \frac{E^b_j}{\gamma}$ (or by E^b_j, but with a factor γ in the RHS of the following Poisson Bracket):

$$\left\{A^i_a\left(x\right), \tilde{E}^b_j\left(x'\right)\right\} = \delta^i_j\delta^b_a\delta^3\left(x - x'\right), \tag{11}$$

The symplectic pair are known as Ashtekar new variables.

The Hamiltonian density is:

$$\mathcal{H} = e^{\mu}_{I} e^{\nu J} F^{I}_{\mu\nu J}(A) - E^{a3}_{i} \dot{A}^{i}_{a} = A^{i}_{0} \mathcal{G}_i + N\mathcal{C} + N^a \mathcal{C}_a \approx 0$$

with the following constraint density in terms of the new variables:

$$\mathcal{G}_i = D_a \tilde{E}^a_i = D_a \tilde{E}^a_i + \varepsilon_{ij}{}^{k} A^j_a \tilde{E}^a_k, \tag{12}$$

$$\mathcal{C}_a = F^i_{ab} \tilde{E}^b_i - \frac{1+\gamma^2}{\gamma} K^i_a \mathcal{G}_i, \tag{13}$$

$$\mathcal{C} = \left(F^j_{ab} - (1+\gamma^2)\, \varepsilon^j{}_{mn} K^m_a K^n_b \right) \frac{\varepsilon_j{}^{kl} \tilde{E}^a_k \tilde{E}^b_l}{\det \tilde{E}} + (1+\gamma^2)\, \mathcal{G}^i \partial_a \frac{\tilde{E}^a_i}{\det \tilde{E}}. \tag{14}$$

For a special condition $\gamma = i$, the constraint greatly simplifies to a compact form as follows:

$$\mathcal{G}_i = d_a \tilde{E}^a_i, \tag{15}$$

$$\mathcal{C}_a = F^i_{ab} \tilde{E}^b_i, \tag{16}$$

$$\mathcal{C} = F^j_{ab} \frac{\varepsilon_j{}^{kl} \tilde{E}^a_k \tilde{E}^b_l}{\det \tilde{E}}. \tag{17}$$

Smearing the constraints as in the previous subsection:

$$G\left(\vec{A}\right) = \Delta_\Sigma\, A^i_0 \mathcal{G}_i q^{12} d^3x, \tag{18}$$

$$C(N) = \Delta_\Sigma\, N\mathcal{C} q^{12} d^3x, \tag{19}$$

$$C\left(\vec{N}\right) = \Delta_\Sigma\, N^a \mathcal{C}_a q^{12} d^3x, \tag{20}$$

one could obtain the following constraints algebra:

$$\left\{G\left(\vec{A}\right), G\left(\vec{A'}\right)\right\} = G\left(\left[A^i_0, A'^i_0\right]\right), \qquad \left\{C(N), C(N')\right\} = C\left(\vec{K}\right) - G\left(A^i_a K^a\right), \tag{21}$$

$$\left\{G\left(\vec{A}\right), C(N')\right\} = 0, \qquad \left\{C(N), C\left(\vec{N'}\right)\right\} = -C\left(L_{\vec{N'}} N\right),$$

$$\left\{G\left(\vec{A}\right), C\left(\vec{N'}\right)\right\} = -G\left(L_{\vec{N'}} \vec{N}\right), \qquad \left\{C\left(\vec{N}\right), C\left(\vec{N'}\right)\right\} = C\left(L_{\vec{N}} \vec{N'}\right),$$

with vector $\vec{K}$ is defined by $K^a = \tilde{E}^a_i \tilde{E}^b_j \delta^{ij} (N\partial_b N' - N'\partial_b N)$. The algebra, although they do not construct a true Lie algebra since RHS of (21) is not a structure constant, is closed. Therefore, by restricting on the constraint hypersurface, where $G\left(\vec{A}\right) \approx 0$, $C\left(\vec{N}\right) \approx 0$, and $C(N) \approx 0$, all the primary constraint commutes with themself on the constraint hypersurface. This means the sets of constraint arising from the Ashtekar formulation of GR are primary first class constraint. As a consequence to this, these constraints generate gauge transformation: Gauss constraint generates SU(2) gauge transformation, diffeomorphism constraint generates evolution in space, Hamiltonian constraint generates evolution in time.

3. Canonical Quantization

Since the set of constraints are primary and first class, one could proceed to the quantization procedure, based rigorously on [7]. The steps are:

(i) To promote Poisson bracket $\{\ldots\}$ to commutator $[\ldots]$.

(ii) To associate every dynamical variable $\mathcal{O}$ with an operator $\hat{\mathcal{O}}$.

(iii) To construct space of states where the the operators act to: the target/representation space.

Journal of Physics: Conference Series **1354** (2019) 012002 doi:10.1088/1742-6596/1354/1/012002

3.1. Quantization via Connection Representation

One could immediately quantized the phase space, the Poisson bracket (11) becomes:

$$\left[A_a^i(x), \tilde{E}_j^b(x')\right] = i\hbar\delta_j^i\delta_a^b\delta^3(x-x').$$

The next step is to promote the fundamental variables to the following operators:

$$\begin{aligned} A_a^i \to \hat{A}_a^i &= A_a^i, \\ E_i^a \to \hat{E}_i^a &= \frac{\delta}{\delta A_a^i}, \end{aligned}$$

$\frac{\delta}{\delta A}$ is the functional derivatives with respect to A_a^i. The last step is to obtain the representation space. Normally, the representation space is the square-integrable space over the configuration space of the system. In our case, the configuration space is the space of connection $\mathcal{A} \simeq \mathfrak{su}(2)\times\bigwedge^1(\Sigma)$, therefore the representation space should be the functional space $C^\infty[\mathcal{A}] \simeq L_2[\mathcal{A}]$. Moreover, one could equip the space with a well-defined inner product and completeness requirement; in this case, the functional space is a Hilbert space.

At this point, several problems arise: the properties of the representation space are not well-defined: $C^\infty[\mathcal{A}] \ni \psi[A]$ has infinite (uncountable) dimensions, the measure and the inner product in this space is unknown. Let us neglect this problem for a while. Nevertheless, there exist a remaining problem when one tries to promote contraints (18)-(20) to an operator: the ordering of triads $\tilde{E}$ and curvature $F(A)$ matters. Let us check what can be obtained from each ordering.

3.1.1. Triads on the right: Wilson Loops. In this ordering, the constraint becomes the following operator:

$$\hat{\mathcal{G}}_i\psi[A] = d_a\frac{\delta}{\delta A_a^i}\psi[A], \tag{22}$$

$$\hat{\mathcal{C}}_a\psi[A] = F_{ab}^i\frac{\delta}{\delta A_b^i}\psi[A], \tag{23}$$

$$\hat{\mathcal{C}}\psi[A] = \varepsilon_j^{\ kl}F_{ab}^j\frac{\delta}{\delta A_a^k}\frac{\delta}{\delta A_b^l}\psi[A]. \tag{24}$$

The Gauss constraint becomes an infinitesimal generator of gauge transformation for $SU(2)$ and the diffeomorphism constraint becomes infinitesimal generator of 3D diffeomorphism. The wavefunctional that satisfies the Gauss constraint needs to be invariant under SU(2) transformation, thus the candidate for the solution is the Wilson loop of an SU(2) holonomy:

$$\psi[A] = W[A,\gamma] = \mathrm{tr}U[A,\gamma], \qquad U[A,\gamma] = \hat{P}\exp\oint_\gamma A.$$

In fact, the Wilson loop is also a solution to the Hamiltonian constraint. On the other hand, (1) $W[A,\gamma]$ is not a solution to the diffeomorphism constraint (23), (2) $W[A,\gamma]$ is not a solution to Hamiltonian constraint if the loops contains kinks or intersections, (3) given a metric operator, the wavefunction represents a space with degenerate metric [4]. Unfortunately, research in this direction might be not really useful. Nevertheless, it is historically important since this gives motivational research to the loop direction.

3.1.2. Triads on the left: Chern-Simon Forms. In this ordering, the constraint becomes the following operator:

$$\hat{\mathcal{G}}_i \psi [A] = \frac{\delta}{\delta A^i_a} d_a \psi [A] , \quad (25)$$

$$\hat{\mathcal{C}}_a \psi [A] = \frac{\delta}{\delta A^i_b} F^i_{ab} \psi [A] , \quad (26)$$

$$\hat{\mathcal{C}} \psi [A] = \varepsilon_j^{\ kl} \frac{\delta}{\delta A^k_a} \frac{\delta}{\delta A^l_b} F^j_{ab} \psi [A] . \quad (27)$$

For a reason which will be clear later, let us modify the Hamiltonian constraint by the existence of the cosmological constant:

$$\hat{\mathcal{C}}_\Lambda \psi [A] = \left(\varepsilon_j^{\ kl} \frac{\delta}{\delta A^k_a} \frac{\delta}{\delta A^l_b} F^j_{ab} - \frac{\Lambda}{6} \varepsilon_{jkl} \varepsilon^{abc} \frac{\delta}{\delta A^j_a} \frac{\delta}{\delta A^k_b} \frac{\delta}{\delta A^l_c} \right) \psi [A] .$$

Chern-Simon Theory. At this point, let us review a different subject: the Yang-Mills field. The action of a Yang-Mills fields is given as follows:

$$S_{YM} = \Delta_{\mathcal{M}} \mathcal{L}_{YM} = \Delta_{\mathcal{M}} \operatorname{tr} (F \wedge \star F) . \quad (28)$$

For a case without source, as an example, in the electromagnetism case, the electromagnetic field satisfy self-duality $F = \star F$, such that the action (28) becomes:

$$S_{YM} = \Delta_{\mathcal{M}} \operatorname{tr} (F \wedge F) = \Delta_{\mathcal{M}} \operatorname{tr} \left(F^2 \right) . \quad (29)$$

It is clear that the action is independent from the use of metric, which is included in the definition of the Hodge-dual. This is favourable in the background independence perspective. In fact, one could generalize the action for arbitrary even $2n$-dimension of $\mathcal{M}$, n integers:

$$S_{CS} = \Delta_{\mathcal{M}} \mathcal{L}_{CS} = \Delta_{\mathcal{M}} \operatorname{tr} (F^n) . \quad (30)$$

The term F^n is known as the nth Chern form, and (30) is the Chern-Simon action. One can prove that $d_D F^n = 0$. Since F^n is closed, it is reasonable to ask if one could obtain a solution to $d_D F^n = 0$, namely a $(2n-1)$-form potential ϕ such that $d_D \phi = F^n$. This is equivalent with requiring the nth Chern form F^n to be exact. An equivalence class of exact F^n is known as the nth Chern class, and the corresponding potential ϕ as the Chern-Simon form. This solution will be important if one is interested to consider the bulk/boundary correspondence of the field. For $n = 4$ case, one could prove that $\phi = A \wedge dA + \frac{2}{3} A \wedge A \wedge A$, a Chern-Simon 3-form. Therefore, using Stokes theorem, one could write (29) as:

$$S_{YM} = \Delta_{\mathcal{M}} \operatorname{tr} (F \wedge F) = \Delta_{\partial \mathcal{M}} \operatorname{tr} \left(A \wedge dA + \frac{2}{3} A \wedge A \wedge A \right) ,$$

with $\partial \mathcal{M}$ is the boundary of $\mathcal{M}$.

Now, let us construct a Chern-Simon state as follows:

$$\psi [A] = e^{-\frac{i6}{\Lambda} S_{YM}[A]} = e^{-\frac{i6}{\Lambda} \Delta_\Sigma \operatorname{tr} \left(A \wedge dA + \frac{2}{3} A \wedge A \wedge A \right)} .$$

It is widely known as Kodama state [11]. Remarkably, it is the solution to these three constraints, but with an existence of a positive cosmological constant. The Kodama state is interpreted as the ground state wavefunction for a deSitter space. Nevertheless, it is not a state for a gravitational field, one could attempt to find a solution based on the action of GR, and the research in this direction is still on progress.

Journal of Physics: Conference Series **1354** (2019) 012002 doi:10.1088/1742-6596/1354/1/012002

3.2. Quantization via Holonomy Representation

It had been explained in the previous chapters that there exist a wavefunction of GR in the form of Kodama state. Nevertheless, the Hilbert space of this state, had not been obtained in a rigorous manner. One could argue that the Ashtekar new variables are not the best phase-space variables to quantize GR; one needs to search for another phase-space variables. Implementing the main idea of the construction of Lattice Gauge Theories, it turns out that the good variables compatible with the quantization procedure are the regularized version of the Ashtekar variables.

3.2.1. Regularization. The 3-connections A, which is an $\mathfrak{su}(2)$-valued 1-form, is smeared along a curve (line, or link) l, and is known as *holonomy*:

$$U_l\left[A,\gamma\right] = \hat{P}\exp \Delta_l\ {}^3A. \tag{31}$$

This is also a solution to the following parallel transport equation along curve γ, with $\gamma'(\tau) = \frac{d\gamma(\tau)}{d\tau}$:

$$D_{\gamma'(\tau)}U\left[A,\gamma(\tau)\right] = \frac{dU\left[A,\gamma(\tau)\right]}{d\tau} + A\left(\gamma'(\tau)\right)U\left[A,\gamma(\tau)\right] = 0.$$

Meanwhile the 'electric' triads $\tilde{E} = \frac{E}{\gamma}$, which is a matrix-valued 2-form, is smeared on a portion of surface area S, known as the *flux*:

$$\tilde{E}_l = \tilde{E}_l\left(S\right) = \Delta_S\,\tilde{E}. \tag{32}$$

The index l in the flux indicates that the (infinitesimal) surface S in 3D hypersurface Σ always posses a link l which is defined as its (lattice) dual, thus it is convenient to label the flux crossing the surface with the index of the link. The smeared phase-space variable associated on link l is now (U_l, E_l), where the U_l is smeared along the link, and $\tilde{E}_l$ is smeared along the surface (lattice)-dual to the link. They satisfy the symplectic structure through the Poisson bracket:

$$\begin{aligned}\left\{U_l, U_{l'}\right\} &= 0,\\ \left\{U_l, \tilde{E}^a_{l'}\right\} &= \delta_{ll'}\bar{\sigma}^a U_l,\\ \left\{\tilde{E}^a_l, \tilde{E}^b_{l'}\right\} &= \delta_{ll'}\varepsilon^{ab}_{\ \ c}\tilde{E}^c_l,\end{aligned}$$

this algebra is known as the holonomy-flux algebra, labeled by $\mathscr{J}_\Gamma$, with Γ is a collection of links and node known as a graph. $\bar{\sigma}^a$ is the Pauli matrices. There are some subleties concerning the orientation of the flux $\tilde{E}$ with respect to the surface S. In this article, we always take the flux to cross surface S positively. For a more subtle derivation consult [3]. Clearly, U_l is an element of $SU(2)$, since it is an 'exponential map' of ${}^3A \in \mathfrak{su}(2)$, while $\tilde{E}^a_l$, which are components of $\tilde{E}_l$, from (35), is an element of $\mathfrak{su}(2)$, since they satisfy the $\mathfrak{su}(2)$ algebra. Therefore, these variables are conjugate to each other, forming an elements of 'regularized' phase space $(U_l, J^a_l) \in T^*SU(2) \simeq SU(2) \times \mathfrak{su}(2)^*$ (since $\mathfrak{su}(2) \sim \mathfrak{su}(2)^*$).

The smearing used for this regularization is different from the standard smearing use in QFT, namely, the smearing over a portion of a 3D volume. This happens because the smearing is defined with respect to a specific background metric. For general relativity, one use different type of smearing such that it does not assume a fix background metric, since the theory is expected to be background independent.

Journal of Physics: Conference Series **1354** (2019) 012002 doi:10.1088/1742-6596/1354/1/012002

3.2.2. Quantization. Repeating the quantization on the regularized variables, the holonomy-flux algebra naturally satisfies the following algebra relation:

$$[U_l, U_{l'}] = 0, \tag{33}$$
$$[U_l, E^a_{l'}] = i\ell^2\gamma\delta_{ll'}\bar{\sigma}^a U_l, \tag{34}$$
$$\left[E^a_l, E^b_{l'}\right] = i\ell^2\gamma\delta_{ll'}\varepsilon^{ab}_{\ \ c}E^c_l. \tag{35}$$

ℓ is a constant equal to $\sqrt{\frac{\hbar G}{c^3}}$, known as *Planck length.* Notice that we use E instead of $\tilde{E}$, which give rise to the Immirzi factor γ on the RHS of (34)-(35). Promoting the phase-space elements as operators, one obtains:

$$U_l \rightarrow \hat{U}_l\psi[A] = U_l\psi[A], \tag{36}$$
$$E_l \rightarrow \hat{E}_l\psi[A] = \vec{J}_l\psi[A], \tag{37}$$

with $\vec{J}_l$ is a left invariant vector field on SU(2) generated by Pauli matrices $\bar{\sigma}^a$ if the orientation of the link points outward from the node and right invariant if it points inward:

$$\vec{J}_l : \mathcal{G} \rightarrow \mathfrak{g}, \qquad \vec{J}_l : g_l \mapsto \vec{J}_l(g_l) = \begin{cases} L^a_l(g_l) = i\frac{d}{d\tau}\left(e^{\bar{\sigma}^a\tau}g_l\right)\Big|_{\tau=0} \\ R^a_l(g_l) = i\frac{d}{d\tau}\left(g_l\, e^{\bar{\sigma}^a\tau}\right)\Big|_{\tau=0} \end{cases}. \tag{38}$$

The flux is defined as an invariant vector field acting on $SU(2)$ with a reason which will be clear later. The phase-space variables are automatically the basic operators of the canonical theory; any operator $\hat{\mathcal{O}}$, including the constraints, can be written in the form of these basic operators.

The last step of quantization is constructing the Hilbert space. Since the regularized phase space variables is now (U, J), the regularized configuration space is $SU(2) \ni U_l$. This made us possible to define the flux operator in (38) as an invariant vector field working on SU(2). Therefore, the representation space of quantum gravity could be build over $SU(2)$, instead of $\mathcal{A}$. To be precise, the representation space is exactly the space of square-integrable function of $SU(2)$: $C^\infty[SU(2)] \simeq L_2[SU(2)]$. This allows us to define the Hilbert space, which will be explained in detail in the next section.

4. Kinematical Part of Loop Quantum Gravity

4.1. Cylindrical Functions

4.1.1. Hilbert space of quantum gravity. As explained in the previous section, the representation space over the connection $C^\infty[\mathcal{A}]$ is problematic, i.e., one can not define an inner product, measure, and completeness, which are possesed by a Hilbert space. Therefore, with the connection representation, one cannot obtain the Hilbert space of the theory. Nevertheless, there is a possible way to obtain the Hilbert space using the holonomy representation.

Let a functional over connection A be written as:

$$\psi[A] = \langle A|\psi\rangle \in C^\infty[\mathcal{A}].$$

Now let us consider a subset of $C^\infty[\mathcal{A}]$, say Cyl$[\mathcal{A}]$, such that each element is a function of the holonomy $U = U[\gamma, A]$, namely, the dependence on A enters through U. Such functional which are build from some subset of the field are generally known as cylindrical functions. We label this kind of function as:

$$\psi_{\Gamma,\varphi}[A] = \langle A|\psi_{\Gamma,\varphi}\rangle = \varphi(U_{l_1}[A], .., U_{l_L}[A]) \in \text{Cyl}[\mathcal{A}] \simeq C^\infty[SU(2)] \subset C^\infty[\mathcal{A}].$$

Γ is a graph constructed from intersecting loops, where each link l contains the information of the holonomy U_l. Notice that since $\mathcal{A} \sim \mathfrak{su}(2) \times \bigwedge^1(\Sigma)$, then the space of holonomies, new configuration space, is clearly $SU(2)$ which is compact. Because of this, one has completeness, and could define the following inner product on $\mathrm{Cyl}[\mathcal{A}] \simeq C^\infty[SU(2)]$:

$$\langle \psi_{\Gamma,\varphi} | \psi_{\Gamma,\varphi'} \rangle = \underset{SU(2)}{\Delta} \prod_l \mathrm{d}U_l\, \varphi^*\left(U_{l_1}[A], .., U_{l_L}[A]\right) \varphi'\left(U_{l_1}[A], .., U_{l_L}[A]\right),$$

with $\mathrm{d}U_l$ is the Haar measure on $SU(2)$. Moreover, Ashtekar and Lewandowski shows that $\mathrm{Cyl}[\mathcal{A}]$ could be extended to define a Hilbert space over a connection, $\mathcal{H}_{AL} \simeq L_2[\mathcal{A}, \mathrm{d}\mu_{AL}]$, using the Ashtekar-Lewandowski measure $\mathrm{d}\mu_{AL}$ [12, 13]. Therefore, one obtains a candidate of Hilbert space for quantum gravity.

4.1.2. The basis on Ashtekar-Lewandowski Hilbert space. Since now the cylindrical function $|\psi_{\Gamma,\varphi}\rangle$ is an element of Hilbert space $\mathcal{H}_{AL}$, one can use Peter-Weyl theorem: *A basis in* $\mathcal{H} \simeq L_2[\mathcal{G}, \mathrm{d}\mu_{\mathrm{Haar}}]$, *that is, a space of function over a compact group* $\mathcal{G}$, *is given by the matrix element of unitary, irreducible representations of* $\mathcal{G}$. In the LQG case, $\mathcal{G} \simeq SU(2)$, therefore, one could obtain the basis of $\mathcal{H}_{AL} \simeq L_2[\mathcal{A}, \mathrm{d}\mu_{AL}]$ in the following steps.

Firstly, let us construct the Hilbert space for a single link of graph Γ. A link $l = (n, n')$ connecting node n to n' of the graph is equipped by an element of phase space on the boundary, $(U_l, J_l) \in T^*SU(2)$, together with the label of spin-j, the irreducible representation of $SU(2)$ in $(2j+1)$-dimension. The representation space for each link is build over $SU(2)$, and from the theorem one has:

$$\mathcal{H}_l = L_2\left[SU(2), \mathrm{d}\mu_{\mathrm{Haar}}\right] \simeq \bigoplus^{j_l} \left(\mathcal{H}_{j_l} \otimes \mathcal{H}^*_{j_l}\right), \qquad \mathcal{H}_{j_l} \simeq L^2\left[\mathbb{R}^n, \mathrm{d}\mu\right]. \tag{39}$$

Thus, the basis in $\mathcal{H}_l \simeq L_2[SU(2)]$ is:

$$|j_l, m_l, n_l\rangle = |j_l, m_l\rangle \langle j_l, n_l| \in \mathcal{H}_l \simeq \bigoplus^{j_l} \left(\mathcal{H}_{j_l} \otimes \mathcal{H}^*_{j_l}\right), \tag{40}$$

with each spin basis is related to one end of the link, see Figure. 1.

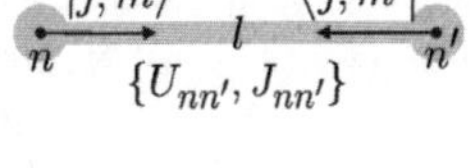

Figure 1. Variables and spin quantum numbers attached to one link of a graph.

For a graph containing n-links, the Hilbert space is the product of (39), namely:

$$\mathcal{H}_\Gamma = L_2\left[SU(2)^{|\mathcal{L}|}, \mathrm{d}\mu_{\mathrm{Haar}}\right] = \bigotimes_l^{|\mathcal{L}|} \mathcal{H}_l = \bigotimes_l^{|\mathcal{L}|} \bigoplus^{j_l} \left(\mathcal{H}_{j_l} \otimes \mathcal{H}^*_{j_l}\right), \tag{41}$$

and the basis, which could be represented with n non-connected strains, is:

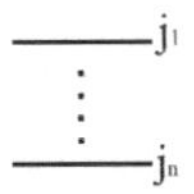

$$= \bigotimes_{l=1}^{n} |j_l, m_l, n_l\rangle$$

$\otimes |j_l, m_l, n_l\rangle$ labels the basis of matrix representation of the wavefunctional.

4.1.3. General states. Let us write the vector state $|\psi_{\Gamma,\varphi}\rangle \in \mathcal{H}_{AL}$ of an arbitrary graph Γ having $|\mathcal{N}|$ nodes and $|\mathcal{L}|$ links, in the representation of holonomy as follows:

$$|\varphi\rangle = \underset{SU(2)}{\Delta} \prod_l \mathrm{d}U_l \left\langle \otimes^l U_{l_1}[A] |\varphi\rangle \left| \otimes^l U_{l_1}[A] \right\rangle = \underset{SU(2)}{\Delta} \prod_l \mathrm{d}U_l \varphi\left(U_{l_1}[A], .., U_{l_L}[A]\right) \left| \otimes^l U_{l_1}[A] \right\rangle,$$

where we have $\varphi\left(U_{l_1}[A], .., U_{l_L}[A]\right) = \langle \otimes^l U_{l_1}[A] |\varphi\rangle$ which is a function over SU(2). Using the completeness and orthogonality of (40), one could write $\varphi\left(U_{l_1}[A], .., U_{l_L}[A]\right)$ as a linear combination of irreducible representation of SU(2), as follows:

$$\varphi\left(U_{l_1}[A], .., U_{l_L}[A]\right) = \sum_{j_l, m_l, n_l} \left\langle \otimes^l U_{l_1}[A] | \otimes_l j_l, m_l, n_l \right\rangle \left\langle \otimes_l j_l, m_l, n_l | \varphi \right\rangle = \sum_{j_l, m_l, n_l} \varphi^{j_l}_{m_l n_l} \prod_l D^{j_l}_{m_l n_l}(U_l), \tag{42}$$

with $D^{j_l}_{m_l n_l}(U_l) = \langle U_l[A] \mid j_l, m_l, n_l\rangle$, is the component of the *Wigner-D matrix*, or the *rotator.*

4.2. Gauss Constraint and the Kinematical Hilbert Space

4.2.1. Gauss constraint in flux-representation: closure constraint of a quantum polyhedron. Having the well-defined Ashtekar-Lewandowski Hilbert space and its complete basis, one could start to promote the constraints as operators and find the wavefunctional solving the constraints. In this review, we only consider the Gauss constraint. The Gauss constraint (12) is a divergence theorem, it can be interpreted as the 'incompressibility' of the flux E^a_i. We can write the Gauss constraint without indices as follows:

$$\mathcal{G} = d_\Sigma E = 0. \tag{43}$$

with d_Σ is the exterior-covariant derivative on the spatial slice Σ. The quantity (43) is a 3-form, therefore, to define the regularization, one needs to smeared the quantity over a single finite region with volume R_i, and by using Stokes theorem, one obtains:

$$\Delta_{R_i} d_\Sigma E = \Delta_{\partial R_i} E = 0,$$

with ∂R is the closed surface enclosing the finite volume R_i. Using the smeared variable of the infinitesimal momentum over a finite surface, one obtains:

$$\Delta_{\partial R_i} E = \lim_{\mathbf{a}_l \to 0} \sum_{\mathbf{a}_l \subset \partial R_i} E_l(S).$$

Defining a regulator or a cut-off which prevent $\mathbf{a}_l \to 0$, the regularized Gauss constraint is obtained as follows:

$$\mathcal{G}_\Delta = \sum_{\mathbf{a}_l \subset \partial R} E_l(S).$$

By promoting the Gauss constraint to an operator using (37), one has:

$$\mathcal{G}_\Delta \to \hat{G}|\varphi\rangle = \sum_l \hat{E}_l(S)|\varphi\rangle = \sum_l \hat{J}_l|\varphi\rangle = 0, \tag{44}$$

with the index l labels the link describing the area $\mathbf{a}_l$. (44) can be interpreted as the quantum operator version of the classical closure constraint of a polyhedron, where flux J_l correspond to the area $\mathbf{a}_l$ of the polygons enclosing the polyhedron, see Figure 2. For now, let us choose for

Journal of Physics: Conference Series **1354** (2019) 012002 doi:10.1088/1742-6596/1354/1/012002

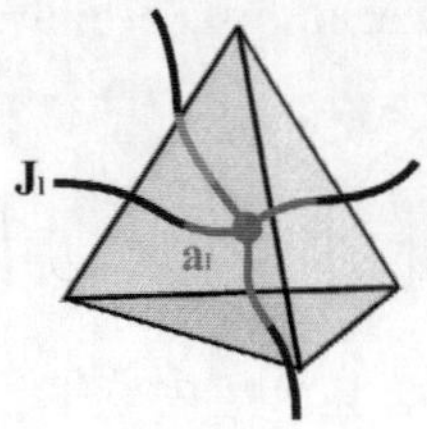

Figure 2. A quantum tetrahedron.

simplicity an $|\mathcal{L}|$-valent graph, that is, a graph containing only a single node n with $|\mathcal{L}|$-links $l = 1, .., |\mathcal{L}|$, pointing outward from n. (44) clearly becomes:

$$\sum_{l=1}^{n} J_l |\varphi\rangle = J_{1..n} |\varphi\rangle = 0. \tag{45}$$

The eigenvalue equation of (45) is translated as:

$$\begin{aligned} |J_{1..n}| |\varphi\rangle &= 0, \\ J_{1..n}^{(z)} |\varphi\rangle &= 0, \end{aligned} \tag{46}$$

requiring the total spin and magnetic quantum number $j_{1...n} = 0$ and $m_{1..n} = 0$ (in fact, $m_{1..n}$ is automatically zero if the first relation is satisfied, since $-j \leq m \leq j$). The state $|\varphi\rangle \in \mathcal{H}_\Gamma$ satisfying (46) automatically satisfies the quantum Gauss constraint. The *kinematical Hilbert space*:

$$\mathcal{K}_\Gamma \subset \mathcal{H}_\Gamma, \qquad \mathcal{K}_\Gamma \sim L_2 \left[SU(2)^{|\mathcal{L}|} / SU(2)^{|\mathcal{N}|} \right]$$

is an invariant subspace of $\mathcal{H}$ which satisfy the quantum Gauss constraint on each node, and the basis which spans $\mathcal{K}$ is called as the *spin network* basis, which will be discussed in detail in the next subsections.

4.2.2. Gauss constraint in holonomy basis: group averaging procedure. The construction of quantum Gauss operator in the previous subsection provides a clear geometrical interpretation. Nevertheless, a formal derivation on states satisfying the Gauss constraint use the procedure borrowed from Lattice Gauge Theory, namely, the group averaging procedure. From (15), it is known that the Gauss constraint is the (infinitesimal) generator of SU(2) gauge transformation.

The finite gauge transformation acts on the holonomy in the following way:

$$U_l [A] \to U_l' [A] = h_l (\gamma (s)) U_l [A] h_l^{-1} (\gamma(t)), \tag{47}$$

where *source* s is the origin of curve γ (or link l), and *target* t is the endpoint of curve γ (s and t are clearly nodes of the graph Γ). In the matrix representation, (47) acts as:

$$D_{mn}^{(j)} (U_l [A]) \to D_{mn}^{(j)} \left(U_l' [A] \right) = D_{ma}^{(j)} (h_l (\gamma (s))) D_{b}^{(j)a} (U_l [A]) D_{n}^{(j)b} (h_l^{-1} (\gamma(t))) .$$

Any wavefunction $\varphi (U [A]) \in \mathcal{H}$ can be written as a linear combination of products of irreducible matrix representation of SU(2) as in (42). One would like to obtain a set of wavefunction $\varphi_{\text{inv}} (U [A]) \in \mathcal{K} \subset \mathcal{H}$, such that the following condition holds:

$$\varphi (U_{l_1} [A] , .., U_{l_L} [A]) \equiv \varphi \left(h_{l_1} (\gamma (s)) U_{l_1} [A] h_{l_1}^{-1} (\gamma(t)) , .., h_{l_L} (\gamma (s)) U_{l_L} [A] h_{l_L}^{-1} (\gamma(t)) \right),$$

That is, the wavefunction needs to be invariant under finite SU(2) gauge transformation. Firstly, one has:

$$\varphi_{\text{inv}}\left(U_{l_1}[A],..,U_{l_L}[A]\right)=\sum_{j_l,m_l,n_l}\varphi^{j_l}_{m_l n_l}\left(\prod_l D^{j_l}_{m_l n_l}(U_l)\right)_{\text{inv}}. \tag{48}$$

Let us focus to the last term containing product of the rotation matrices. Let us collect the product of matrices by the nodes of graph Γ as follows:

$$\prod_l D^{(j_l)}_{m_l n_l}\left(U_l[A]\right)=\prod_{n\in\Gamma}\prod_{\left(\substack{l\in\Gamma,\\ n\cap l\neq\tilde{A}}\right)} D^{(j_l)}_{m_l n_l}\left(U_l[A]\right).$$

The gauge transformation only acts on the nodes, so let us concentrate on a single node where all of the links are pointing outward from node n. The gauge transformation at source node s acts on the *half* of the rotation matrix, i.e on the first index; this is clearly the left action by gauge group $h_l(\gamma(s))$:

$$\left.\prod_l D^{(j_l)}_{m_l n_l}\left(U'_l[A]\right)\right|_{n_t}=\prod_{\left(\substack{l\in\Gamma,\\ n\cap l\neq\tilde{A}}\right)} D^{(j_l)}_{m_l a}\left(h_l(\gamma(s))\right)D^{(j_l)a}_{\ \ n_l}\left(U_l[A]\right).$$

The invariance on the node n means the following condition is satisfied:

$$\left.\prod_l D^{(j_l)}_{m_l n_l}\left(U'_l[A]\right)\right|_{n}=\prod_{\left(\substack{l\in\Gamma,\\ n\cap l\neq\tilde{A}}\right)} D^{(j_l)}_{m_l a}\left(h_l(\gamma(s))\right)D^{(j_l)a}_{\ \ n_l}\left(U_l[A]\right)\equiv\left.\prod_{\left(\substack{l\in\Gamma,\\ n\cap l\neq\tilde{A}}\right)} D^{(j_l)}_{m_l n_l}\left(U_l[A]\right)\right|_{n}. \tag{49}$$

The product of matrices satisfying this property can be obtained by group averaging procedure, or averaging the matrices by a measure i such that condition (49) is satisfied:

$$\left(\left.\prod_l D^{(j_l)}_{m_l n_l}\left(U_l[A]\right)\right|_{n}\right)_{\text{inv}}=\left(\left.\prod_{\left(\substack{l\in\Gamma,\\ n\cap l\neq\tilde{A}}\right)} D^{(j_l)}_{m_l n_l}\left(U_l[A]\right)\right|_{n}\right)_{\text{inv}}=\sum_{\alpha_l,\beta_l} i^{\alpha_1\beta_1..\alpha_l\beta_l}_{(n)m_l n_l}\left.\prod_{\left(\substack{l\in\Gamma,\\ n\cap l\neq\tilde{A}}\right)} D^{(j_l)}_{\alpha_l\beta_l}\left(U_l[A]\right)\right|_{n}.$$

One could define the gauge transformation at target node t acting on the second index; which is the right action $h_l^{-1}(\gamma(t))$ in a similar way. Now taking all the nodes on the graph Γ, one obtains:

$$\left(\prod_l D^{j_l}_{m_l n_l}(U_l)\right)_{\text{inv}}=\prod_{n\in\Gamma}\left(\left.\prod_l D^{(j_l)}_{m_l n_l}\left(U_l[A]\right)\right|_{n}\right)_{\text{inv}}=\sum_{\alpha_l,\beta_l}\prod_{n\in\Gamma} i^{\alpha_1\beta_1..\alpha_l\beta_l}_{(n)m_l n_l}\left.\prod_{\left(\substack{l\in\Gamma,\\ n\cap l\neq\tilde{A}}\right)} D^{(j_l)}_{\alpha_l\beta_l}\left(U_l[A]\right)\right|_{n}, \tag{50}$$

and moreover inserting (50) to (48) gives:

$$\varphi_{\text{inv}}\left(U_{l_1}[A],..,U_{l_L}[A]\right)=\sum_{j_l,m_l,n_l}\varphi^{j_l}_{m_l n_l}\sum_{\alpha_l,\beta_l}\prod_{n\in\Gamma} i^{\alpha_1\beta_1..\alpha_l\beta_l}_{(n)m_l n_l}\left.\prod_{\left(\substack{l\in\Gamma,\\ n\cap l\neq\tilde{A}}\right)} D^{(j_l)}_{\alpha_l\beta_l}\left(U_l[A]\right)\right|_{n}. \tag{51}$$

The components of the equivariant map i is known as the *intertwinner*, which could be obtained from a contraction of $\{3j\}$-symbols. Written in a more compact manner, the state satisfying Gauss constraint can be written as:

$$\varphi^{\text{inv}}_{j_l,v_n,\Gamma}\left(U_{l_1}[A],..,U_{l_L}[A]\right)=\left\langle\bigotimes^{n}\boldsymbol{i}_{(n)}\middle|\bigotimes^{l}\boldsymbol{D}^{(j_l)}\right\rangle_{\Gamma}.$$

Journal of Physics: Conference Series **1354** (2019) 012002 doi:10.1088/1742-6596/1354/1/012002

4.3. Lattice-graph and Spin-network states

Constructing spin network states. The graphical interpretation of spin network is more clearer in the basis of matrix representation $|j_l, m_l, n_l\rangle$ instead of the holonomy basis $|U_l\rangle$. The steps for constructing a spin network states can be written in the following, based on [14]:

(i) Having an arbitrary connected graph Γ, with $|\mathcal{N}|$ nodes and $|\mathcal{L}|$ links, construct the Ashtekar-Lewandowski Hilbert space as the product of the Hilbert space of the link by relation (41). The basis functional on this space is $\otimes |j_l, m_l, n_l\rangle$, by Peter-Weyl theorem. We called this the $\otimes$-product basis.

(ii) On each node, collect half of the $|j_l, m_l, n_l\rangle = |j_l, m_l\rangle \langle j_l, n_l|$ basis, that is, either $|j_l, m_l\rangle$ or $\langle j_l, n_l|$ depending on the graph, and transform them to the $\oplus$-sum basis, that is, sum all the half-spins on the node. The Hilbert space satisfies $\otimes^j \mathcal{H}_j = \oplus_{j_{\min}}^{j_{\max}} \mathcal{H}$. The transformation matrix components are the Clebsh-Gordan coefficients, or more general, the recoupling coefficient i_{CG}. This $\oplus$-sum basis equally spans the Hilbert space (41).

(iii) Implement the Gauss constraint on each node, that is, by requiring the total spins on each node to satisfy (46). This procedure selects only the singlet spaces $\mathcal{H}_{j_{\min}} = \mathcal{H}_0$ (if they exist). This is equal with giving a condition to the recoupling coefficient such that they are equal with the intertwiner, i.e., $i_{CG} = i$. The singlet $\oplus$-sum basis is the spin network basis.

(iv) Contract the spin network basis with the dual-basis in holonomy representation $\langle U_l|$ to obtain the rotation matrices as in (42). One obtains the basis of spin network in terms of (50). Any general spin network states are a linear combination of the spin network basis, as in (51).

Example: 4-valent graph. Taking the invariant subspace of the spin network is implementing the quantum Gauss constraint on each node. In the $\oplus$-basis, implementing this constraint will set the quantum number $j_{1234} = 0$, which is the Clebsch-Gordan condition. This will cause $m_{1234} = 0$, and $j_{123} = j_4$ for the 4-valent graph case:

$$\begin{aligned}\pi_n \left| j_{1234}, m_{1234}, j_{123}, j_{12}, \prod_{l=1}^{n=4} (j_l, m_l) \right\rangle &= \left| 0, 0, j_{123} = j_4, j_{12}, \prod_{l=1}^{n=4} (j_l, m_l) \right\rangle, \\ &= \left| j_{12}, \prod_{l=1}^{n=4} (j_l, m_l) \right\rangle = |j_l, i_\alpha\rangle,\end{aligned}$$

with $|j_l, i_\alpha\rangle$ is the basis of $\mathcal{K}_\gamma$ for 4-valent graph. Written diagramatically:

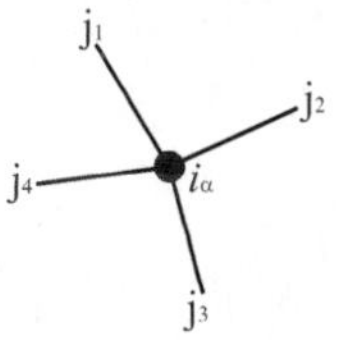

$$= \quad |j_1, .., j_4, i_\alpha\rangle$$
,

with the full-coloured node labels the gauge-invariant basis.

4.4. The geometric operators and quanta of space

The spin network state describes the state of the graph, which is embedded on the spatial hypersurface Σ. It contains the informations of the hypersurface. Since the graph has finite nodes and links, it describes a discretized space, where the nodes correspond to a polyhedra, and the links correspond to flat surfaces enclosing each polyhedron, see Figure 2. Therefore, one could construct the geometric operators such as area and volume operators.

4.4.1. Area operator. The classical area of an arbitrary surface S on Σ is defined as:

$$A(S) = \Delta_S\, ds_1 ds_2 \sqrt{E_i^a E^{bi} n_a n_b}.$$

A regularization gives:

$$A_N(S) = \sum_{I=1}^{N} \sqrt{E_i(S_I)\, E^i(S_I)},$$

with $A(S) = \lim_{N\to\infty} A_N(S)$. Promoting the classical area to operator by (37) gives immediately the area operator of LQG:

$$A_N(S) \to \hat{A}(S), \qquad \hat{A}(S)\,|j,m,n\rangle = \sum_{I=1}^{N} \sqrt{\hat{E}_i(S_I)\, \hat{E}^i(S_I)}\,|j,m,n\rangle\,, \tag{52}$$

acting on $|j,m,n\rangle$. Working on the basis of matrix representation, the spectrum of the area operator are:

$$\hat{A}(S)\,|j,m,n\rangle = \sum_{I=1}^{N} \sqrt{\hat{E}_i(S_I)\, \hat{E}^i(S_I)}\,|j,m,n\rangle = \gamma^2 \ell^2 j\,(j+1)\,|j,m,n\rangle\,,$$

which are discrete.

4.4.2. Volume operator. The classical volume of region R on Σ is defined as:

$$V_R(x) = \Delta_R\, \mathrm{vol}_\Sigma = \Delta_R\, d^3x \sqrt{q(x)},$$

with q is the determinant of metric q_{ij} of Σ. Using the relation between the 3-metric and the triads, one obtains: $q = E$. Therefore:

$$V_R(x) = \Delta_R\, d^3x \sqrt{E(x)} = \Delta_R\, d^3x \sqrt{\left|\frac{1}{3!}\varepsilon^{ijk}\varepsilon_{abc} E_i^a(x)\, E_j^b(x)\, E_k^c(x)\right|},$$

Promoting the classical volume to operator gives:

$$V_R(x) \to \hat{V}_R(x) = \Delta_R\, d^3x \sqrt{\left|\frac{1}{3!}\varepsilon^{ijk}\varepsilon_{abc} \hat{E}_i^a(x)\, \hat{E}_j^b(x)\, \hat{E}_k^c(x)\right|},$$

By a specific regularization defined by [13], the Ashtekar-Lewandowski volume operator is:

$$\hat{V}_R(x) = \ell_P^3 \sum_{v\in V(\Gamma)} \delta^3(x,v)\, V_{v,\Gamma},$$

$$V_{v,\Gamma} = V_{AL} = \sqrt{\left|z \sum_{a>b>c} \epsilon(e_a, e_b, e_c)\, \varepsilon^{ijk} \hat{J}_i^a \hat{J}_j^b \hat{J}_k^c\right|},$$

or

$$V_{v,\Gamma} = V_{AL} = \sqrt{\left|Z \sum_{a,b,c} \epsilon(e_a, e_b, e_c)\, \hat{q}_{abc}\right|},$$

The spectrum of the volume operator is quite complicated to obtain since they are in general, not analytical. For a complete derivation, consult [15]. Nevertheless one can conclude that the volume are discrete. The spin-network graph describe quantum polyhedron, which could be viewed as the quanta of space, having discrete surface areas and volumes and fuzzy shapes.

5. Conclusions

In this article, we have given a brief review on the kinematical part of canonical loop quantum gravity. The kinematical part is complete and well-understood. One of the consequences of LQG coming from the kinematical part is the discreteness of space in the Planck scale, namely the existence of quanta of space. The quanta, having discrete surface areas and volumes, as well as fuzzy shapes, are described by spin-networks: a lattice-graph labeled by spin representations of SU(2). The graph may contain loops, where the SU(2) holonomies describing the intrinsic curvature of the discrete space are located.

The second part of this article will briefly review the dynamical part of canonical loop quantum gravity.

Acknowledgment

S.A. is supported by Institut Teknologi Bandung In House Collaboration Post Doctoral Fellowship.

References

[1] P. Doná, S. Speziale. *Introductory lectures to loop quantum gravity.* (2010). arXiv:gr-qc/1007.0402.
[2] C. Rovelli. *Zakopane lectures on loop gravity.* **PoS** QGQGS2011 **3**. (2011). arXiv:gr-qc/1102.3660.
[3] K. Giesel. *Quantum Geometry.* 100 Years of General Relativity: Vol. **4**: Loop Quantum Gravity. (2017).
[4] J. Pullin. *Knot theory and quantum gravity in loop space: a primer.* AIP Conf. Proc. **317**: 141-190. (1994). arXiv:hep-th/9301028.
[5] A. Palatini. *Deduzione invariantiva delle equazioni gravitazionali dal principio di Hamilton.* Rend. Circ. Mat. Palermo **43**: 203-212. Cosmology and Gravitation. Plenum Press, New York. (1980).
[6] R. Arnowitt, S. Deser, C. Misner. *Dynamical Structure and Definition of Energy in General Relativity.* Phys. Rev. **116** (5): 1322-1330. (1959).
[7] P. A. M. Dirac. *Lectures on Quantum Mechanics.* Snowball Publishing. ISBN-13: 978-1607964322. (2012).
[8] S. Inglima. *Topics on Canonical Gravity. Dissertation.* (2012). https://www.imperial.ac.uk/media
[9] A. Ashtekar. *New Hamiltonian formulation for general relativity.* Phys. Rev. D **36**: 6. (1987). PRD36p1587_1987.pdf.
[10] S. Holst. *Barbero's Hamiltonian derived from a generalized Hilbert-Palatini action.* Phys. Rev. D **53**: 5966–5969. (1996). arXiv:gr-qc/9511026
[11] H. Kodama (1988). *Specialization of Ashtekar's Formalism to Bianchi Cosmology.* Progress of Theoretical Physics. **80** (6): 1024.
[12] A. Ashtekar, J. Lewandowski. *Quantum theory of geometry. I: Area operators.* Class. Quant. Grav. **14**: A55–A82. (1997). arXiv:gr-qc/9602046.
[13] A. Ashtekar, J. Lewandowski. *Quantum theory of geometry. II: Volume operators.* Adv. Theor. Math. Phys. **1**: 388-429. (1998).arXiv:gr-qc/9711031.
[14] J. Brunemann, T. Thiemann. *Simplification of the Spectral Analysis of the Volume Operator in Loop Quantum Gravity.* Class. Quant. Grav. **23**: 1289-1346. (2006) arXiv:gr-qc/0405060
[15] J. Brunemann, D. Rideout. P*roperties of the Volume Operator in Loop Quantum Gravity I: Results.* Class. Quant. Grav. **25**: 065001. (2008). arXiv:gr-qc/0706.0469.

Construction of a one-parameter family of isospectral potential to study resonances in weakly bound halo nuclei

Mahamadun Hasan[1], **Tasrief Surungan**[2] **and Md Abdul Khan**[*3]
[1,3]Department of Physics, Aliah University, IIA/27, Newtown, Kolkata-700160, India
[2]Department of Physics, Hasanuddin University, Makassar-90245, Indonesia
E-mail: *drakhan.rsm.phys@gmail.com;drakhan.phys@aliah.ac.in

Abstract. In this paper, a novel theoretical scheme is presented to investigate resonant levels in weakly bound nuclear systems by the use of isospectral potentials. In this scheme, a new potential is constructed which is strictly isospectral with the original shallow-well potential and has properties which are desirable to make the calculation of resonances more accurate and easier. To demonstrate the effectiveness of the method, the first 0^+ resonance in the neutron-rich ^{22}C is calculated assuming a three-body (^{20}C+n+n) cluster model.

PACS number(s): 21.45.+v, 21.10.Dr, 27.30.+t
Key words: Halo nuclei, resonance, isospectral potential.

1. Introduction

One of the most striking events in physics is the discovery of halo nuclei near the drip-lines after the advancements radioactive ion beam (RIB) facilities. Some of the observed neutron halo nuclei are ^{11}Li, ^{6}He, ^{11}Be, ^{14}Be, ^{17}B [1-4] while proton halo structure is found in ^{8}B [5]. Experimental investigation by Audi et al [6] on $^{18-22}$C inferred a 2n-halo structure of ^{22}C. Structure of halo nuclei is characterized by a stable dense core surrounded by one or more loosely bound nucleon(s) giving rise to an extended tail in the density distribution. This low density tail accounts for the unusually large r.m.s. matter radii (larger than the liquid-drop model prediction of $R_A \propto A^{1/3}$) of such exotic nuclei [7-8]. Further, the low-density tail is supposed to be the consequence of quantum mechanical tunneling of the last nucleon(s) through the combination of weak binding and short-range interaction. Halo nuclei rarely support excited bound states, they merely support one bound state with binding energy typically *less than* 1 MeV. However, these systems may have one or more resonance states above the ground state. Calculations for resonance states generally involve a large degree of computational error. Here we propose an effective technique for calculation of resonances together with ground state using supersymmetric quantum mechanics.

Literature survey reveals that there are three main theoretical approaches to study the structure of 2n-halo nuclei. First one is the microscopic model approach which assumes the valence neutrons to move around the conglomerate of other nucleons (protons and neutrons) without having any stable core. The second one is the cluster or three-body model in which the valence nucleons move around the structureless inert core. And the third one is the microscopic

Content from this work may be used under the terms of the Creative Commons Attribution 3.0 licence. Any further distribution of this work must maintain attribution to the author(s) and the title of the work, journal citation and DOI.
Published under licence by IOP Publishing Ltd

Journal of Physics: Conference Series **1354** (2019) 012003 doi:10.1088/1742-6596/1354/1/012003

cluster model in which the valence nucleons move around the excited or deformed core [9-11]. In most of the models, Jacobi coordinates are used to separate the center of mass motion and the system is described in terms of the relative coordinates.

Study of resonances in few-body system is of particular interest in many areas of physics and especially in weakly bound systems, where only a few bound states are possible. For the calculation of resonant states a number of methods are in use, *viz.,* the positive energy solution of the Faddeev equation [12], complex coordinate rotation (CCR)[13-14], hyperspherical harmonics method (HHM) for scattering states [15], the analytic computation of bound state energies [16], the algebraic version of resonating group method (RGM) [17], continuum-discretized coupled-channels (CDCC) method in a club with cluster-orbital shell model (COSM) [18] etc.

In this work, we present a novel theoretical scheme for calculation of resonant states in weakly bound nuclei. As a prerequisite, it requires the existence of a bound ground state and can handle both of the bound and resonant states on the same footing. The scheme is based on the fact that, for any given potential (say, ω), one can construct a family of isospectral potentials ($\hat{\omega}$), which depends on a free parameter (δ). Whilst the original potential has a skin-deep well followed by a low and excessively wide barrier (poorly supporting the resonant state), the parameter δ can be chosen judiciously to increase the depth of the well and height of the barrier in $\hat{w}$ simultaneously. This deep well and high barrier combination in $\hat{\omega}$ facilitates the calculation of resonant state accurately at the same energy as that in ω, since $\hat{\omega}$ and ω are strictly isospectral.

To check the validity of the scheme, we apply it to investigate the first 0^+ resonant state of the weakly bound ^{22}C nucleus [18]. We assume a three-body cluster model for ^{22}C (2n+^{20}C) in which the valence neutrons are assumed to move around the relatively stable and inert ^{20}C core. The three-body effective potential shows a shallow well with a wide skinny barrier causing a large resonance width. In principle, one can find quasi-bound states in such a shallow potential, but it poses a difficult numerical task. For a finite barrier height, a particle can temporarily be trapped inside the shallow well when its energy is close to the resonance energy. However, there is a finite probability that the particle may tunnel out through the barrier. Thus, an accurate calculation of resonance energy is easily masked by the large resonance width resulting from a large tunneling probability due to a small barrier height. Hence, a straightforward calculation of resonance energy of these systems fails to produce accurate results.

We adopt Jacobi coordintes to separate the centre of mass motion and the hyperspherical harmonics expansion (HHE) method [19] to solve the three-body Schrödinger equation in relative coordinates. The three-body relative wavefunction is expanded in a complete set of hyperspherical harmonics (HH), substitution of which in the Schrödinger equation and use of orthonormality of HH leads to set of coupled differential equations (CDE). The HHE method adoted here is an *essentially exact*, as it involves no other approximation than an eventual truncation of the expansion basis. This truncation of the expansion basis can be done to achieve any desired precision in energy finally restricted by the capacity of the available computer. Hyperspherical convergence theorem [20] permits extrapolation of the truncated set of results to predict the fully converged solution. The theorem found useful in yielding high precision results in atomic systems [21-22] and in nuclear systems [23]. Convergence of HH being generally slow, one needs a large number of CDE's to be solved. Hence, for further simplification of the numerical task, we adopt the hyperspherical adiabatic approximation (HAA) [24] to solve the resulting single differential equation (SDE) involving a three-body effective potential $\omega_0(\rho)$ for the ground state energy, E_0 and corresponding wavefunction $\psi_0(\rho)$ [25].

We next construct the isospectral potential $\hat{\omega}_0(\delta;\rho)$ following prescription of the SSQM [26-28]. Finally we solve the SDE for $\hat{\omega}_0(\delta;\rho)$ for different positive energies. Thereafter we calculate probability of finding the particle trapped in the deep narrow well surrounded by an amplified barrier. A plot of this probability as a function of the energy shows a sharp peak at the resonance energy. The actual width of the resonance is obtained by back-transforming the wave function

$\hat{\psi}_0$ of $\hat{\omega}_0(\delta;\rho)$ to ψ_0 of $\omega_0(\rho)$.

The paper is organized as follows. In sections 2 we briefly review the HHE method and in section 3 we present very short details of the SSQM procedure to construct the one-parameter family of isospectral potential. Results of present calculation are discussed in section 4 and conclusions are drawn in section 5.

2. Hyperspherical Harmonics Expansion Method

The nucleus ^{22}C is treated as a three-body system [18] where the core ^{20}C is labeled as particle 1 and the two valence neutrons as particles 2 and 3 respectively. A set of Jacobi coordinates for a given partition i (in which particle i is a spectator) is defined as :

$$\vec{\xi_i} = a_i(\vec{r_j} - \vec{r_k}); \vec{\eta_i} = \frac{1}{a_i}\left(\vec{r_i} - \frac{m_j\vec{r_j} + m_k\vec{r_k}}{m_j + m_k}\right); \vec{R} = \frac{(m_i\vec{r_i} + m_j\vec{r_j} + m_k\vec{r_k})}{M} \tag{1}$$

where $a_i = \left[\frac{m_j m_k M}{m_i(m_j+m_k)^2}\right]^{\frac{1}{4}}$; $m_i, \vec{r_i}$ are the mass and position of the ith particle and $M = m_i+m_j+m_k$, $\vec{R}$ are those of the centre of mass (CM) of the system. Since the interactions among constituent particles depend on relative separations only, the centre of mass motion separates out and the relative motion of the three-body system is described by:

$$\left[-\frac{\hbar^2}{2\mu}(\nabla^2_{\xi_i} + \nabla^2_{\eta_i}) + V_{jk}(\vec{\xi_i}) + V_{ki}(\vec{\xi_i},\vec{\eta_i}) + V_{ij}(\vec{\xi_i},\vec{\eta_i}) - E\right]\Psi(\vec{\xi_i},\vec{\eta_i}) = 0 \tag{2}$$

where $\mu = \left[\frac{m_i m_j m_k}{M}\right]^{\frac{1}{2}}$ $\rightarrow$ effective mass parameter, V_{ij} $\rightarrow$ interaction potential between particles i and j, $\xi_i = \rho\cos\phi_i$; $\eta_i = \rho\sin\phi_i$; $\phi_i = \tan^{-1}(\frac{\eta_i}{\xi_i})$; $\rho = \sqrt{\xi_i^2 + \eta_i^2}$. The hyperradius ρ, an invariant under three dimensional rotations and permutations of particle indices together with the five angular variables $\Omega_i \rightarrow \{\phi_i, \theta_{x_i}, \phi_{x_i}, \theta_{y_i}, \phi_{y_i}\}$ constitute hyperspherical variables of the system. It is to be noted that hyperangles Ω_i depend on the choice of the particular partition i. In terms of hyperspherical variables (ρ, Ω_i) the three-body Schrödinger equation becomes

$$\left[-\frac{\hbar^2}{2\mu}\{\frac{1}{\rho^5}\frac{\partial}{\partial\rho}(\rho^5\frac{\partial}{\partial\rho}) - \frac{\hat{\mathcal{N}}^2(\Omega_i)}{\rho^2}\} + V(\rho,\Omega_i) - E\right]\Psi(\rho,\Omega_i) = 0 \tag{3}$$

where $V(\rho,\Omega_i) = V_{jk} + V_{ki} + V_{ij}$ is the total interaction potential in the partition i and $\hat{\mathcal{N}}^2(\Omega_i)$ is the square of hyper angular momentum operator satisfying the eigenvalue equation

$$\hat{\mathcal{N}}^2(\Omega_i)\mathcal{Y}_{N\alpha_i}(\Omega_i) = N(N+4)\mathcal{Y}_{N\alpha_i}(\Omega_i) \tag{4}$$

The quantum number N represents the hyperangular momentum and $\alpha_i \equiv \{l_{x_i}, l_{y_i}, L, M\}$ is a short hand notation, where L and M stand for the total orbital angular momentum and its projection respectively. The partition dependent normalized eigenfunctions $\mathcal{Y}_{N\alpha_i}(\Omega_i)$ are called the hyperspherical harmonics (HH), a closed analytic expressions for which can be seen in ref. [12].

In the HHE method, the wavefunction $\Psi(\rho,\Omega_i)$ is expanded in the complete set of hyperspherical harmonics associated with a given partition "i":

$$\Psi(\rho,\Omega_i) = \sum_{N\alpha_i}\frac{U_{N\alpha_i}(\rho)}{\rho^{5/2}}\mathcal{Y}_{N\alpha_i}(\Omega_i) \tag{5}$$

Substitution of Eq.(5) in Eq.(3) and use of orthonormality of HH, give rise to a set of coupled differential equations (CDE) in ρ:

$$\left.\begin{array}{l}\left[-\frac{\hbar}{2\mu}\frac{d^2}{d\rho^2}+\frac{(N+\frac{3}{2})(N+\frac{5}{2})\hbar^2}{2\mu\rho^2}-E\right]U_{N\alpha_i}(\rho)\\ +\sum_{N'\alpha_i'} < N\alpha_i|V(\rho,\Omega_i)|N'\alpha_i' > U_{N'\alpha_i'}(\rho)=0.\end{array}\right\} \quad (6)$$

in which

$$< N\alpha_i|V|N'\alpha_i' >= \int \mathcal{Y}^*_{N\alpha_i}(\Omega_i)V(\rho,\Omega_i)\mathcal{Y}_{N'\alpha_i'}(\Omega_i)d\Omega_i \quad (7)$$

The set of CDE's, Eq.(6), is in principle an infinite one which arises due to an infinite number of basis states. For practical purposes, the HH expansion in Eq.(5) is truncated by retaining all N values upto a maximum of N_{max}. For a given N, all allowed values of α_i are included. The number of basis states is further restricted by symmetry requirements and associated conserved quantum numbers. The reduced set of CDE's are solved adopting hyperspherical adiabatic approximation (HAA) [24]. In HAA the CDE's are approximated by a single differential equation assuming that the hyperradial motion is much slower compared to hyperangular motion. Hence the angular motion is first solved for a fixed value of ρ. This corresponds to diagonalization of the potential matrix (including the hyper centrifugal repulsion term) for each ρ-mesh point and choosing the lowest eigenvalue $\omega_0(\rho)$ as the lowest eigen potential [25]. Then the energy of the system is obtained by solving the hyperradial motion for the chosen lowest eigen potential ($\omega_0(\rho)$), which is the effective potential for the hyperradial motion:

$$\left[-\frac{\hbar}{2\mu}\frac{d^2}{d\rho^2}+\omega_0(\rho)-E\right]\psi_0(\rho)=0 \quad (8)$$

Renormalized Numerov method algorithm subject to appropriate boundary conditions in the limit $\rho\to 0$ and $\rho\to\infty$ is then applied to solve Eq.(8) for E ($\leq E_0$). The hyper-partial wave $U_{N\alpha_i}(\rho)$ is given by

$$U_{N\alpha_i}(\rho)=\psi_0(\rho)\chi_{N\alpha_i,0}(\rho) \quad (9)$$

where $\chi_{N\alpha_i,0}(\rho)$ is the $(N\alpha_i)^{th}$ element of the eigenvector, corresponding to the lowest eigen potential $\omega_0(\rho)$.

3. Isospectral Potential

In this section we present a brief review of the scheme of construction of one parameter family of isospectral potentials. In 1-D supersymmetric quantum mechanics, one defines a superpotential for a system in terms of its ground state wave function (ψ_0) [26] as

$$W(\rho)=-\frac{\hbar}{\sqrt{2m}}\frac{\psi_0'(\rho)}{\psi_0(\rho)}. \quad (10)$$

The energy scale is next shifted by the ground state energy (E_0) of the potential $\omega_0(\rho)$, so that in this shifted energy scale the new potential becomes

$$\omega_1(\rho)=\omega_0(\rho)-E_0=\frac{\hbar}{2m}\frac{\psi_0''}{\psi_0} \quad (11)$$

having its ground state at zero energy (i.e. $E_0^{(1)}=0$). It can easily be checked that $\omega_1(\rho)$ is epressible in terms of the superpotential via the Riccati equation

$$\omega_1(\rho)=W^2(\rho)-\frac{\hbar}{\sqrt{2m}}W'(\rho) \quad (12)$$

and the Hamiltonian factorizes as

$$H_1 = -\frac{\hbar}{2m}\frac{d^2}{d\rho^2} + \omega_1(\rho) = A^\dagger A \tag{13}$$

where

$$\left.\begin{aligned} A^\dagger &= W(\rho) - \frac{\hbar}{\sqrt{2m}}\frac{d}{d\rho} \\ A &= W(\rho) + \frac{\hbar}{\sqrt{2m}}\frac{d}{d\rho} \end{aligned}\right\} \tag{14}$$

which are the operators for creation and annihilation of nodes in the wave function. Next we introduce a partner Hamiltonian $H_2 \equiv AA^\dagger$, corresponding to a new partner potential $\omega_2(\rho)$:

$$H_2 = -\frac{\hbar}{2m}\frac{d^2}{d\rho^2} + \omega_2(\rho) = AA^\dagger; \tag{15}$$

$$\omega_2(\rho) = W^2(\rho) + \frac{\hbar}{\sqrt{2m}} W'(\rho) \tag{16}$$

In the above, functions $\omega_1(\rho)$ and $\omega_2(\rho)$ are known as supersymmetric partner potentials.

Now we shall see that the energy eigenvalues and the wave functions of H_1 and H_2 are related. It can be noticed that the energy eigenvalues of both H_1 and H_2 are positive semi-definite ($E_n^{(1,2)} \geq 0$). For $n > 0$, the Schrödinger equation for H_1

$$\left.\begin{aligned} H_1\psi_n^{(1)} &= A^\dagger A\psi_n^{(1)} = E_n^{(1)}\psi_n^{(1)} \\ and, H_2(A\psi_n^{(1)}) &= AA^\dagger A\psi_n^{(1)} = E_n^{(1)}(A\psi_n^{(1)}) \end{aligned}\right\} \tag{17}$$

Similarly,the Schrödinger equation for H_2

$$\left.\begin{aligned} H_2\psi_n^{(2)} &= AA^\dagger\psi_n^{(2)} = E_n^{(2)}\psi_n^{(2)} \\ and, H_1(A^\dagger\psi_n^{(2)}) &= A^\dagger AA^\dagger\psi_n^{(2)} = E_n^{(2)}(A^\dagger\psi_n^{(2)}) \end{aligned}\right\} \tag{18}$$

From Eqs. (17)-(18) and the fact that $E_0^{(1)} = 0$, it is easy understood that the eigenvalues and eigenfunctions of the partner Hamiltonians H_1 and H_2 are related by (n = 0, 1, 2,3, ...)

$$E_n^{(2)} = E_{n+1}^{(1)}, \; E_0^{(1)} = 0, \tag{19}$$

$$\psi_n^{(2)} = \frac{1}{\sqrt{E_{n+1}^{(1)}}}(A\psi_{n+1}^{(1)}), \tag{20}$$

$$\psi_{n+1}^{(1)} = \frac{1}{\sqrt{E_n^{(2)}}}(A^\dagger\psi_n^{(2)}). \tag{21}$$

In the above equations $E_n^{(i)}$ represents the energy of the nth excited state of H_i (i=1,2). It is to be noted that if $\psi_{n+1}^{(1)}(\psi_n^{(2)})$ of $H_1(H_2)$ is normalized then the wave function $\psi_n^{(2)}(\psi_{n+1}^{(1)})$ in Eqs. (20) and (21) is also normalized. Further, the operator $A(A^\dagger)$ not only converts an eigenfunction of $H_1(H_2)$ into an eigenfunction of $H_2(H_1)$ with the same energy, but it also **destroys (creates)** an extra node in the eigenfunction. Since the ground state wave function of H_1 is annihilated by the operator A, this state has no SUSY partner. Thus the scenary we get is that knowning all the eigenfunctions of H_1 we can determine the eigenfunctions of H_2 using the operator A, and vice versa using $A^\dagger$ we can reconstruct all the eigenfunctions of H_1 from those of H_2 except for the ground state.

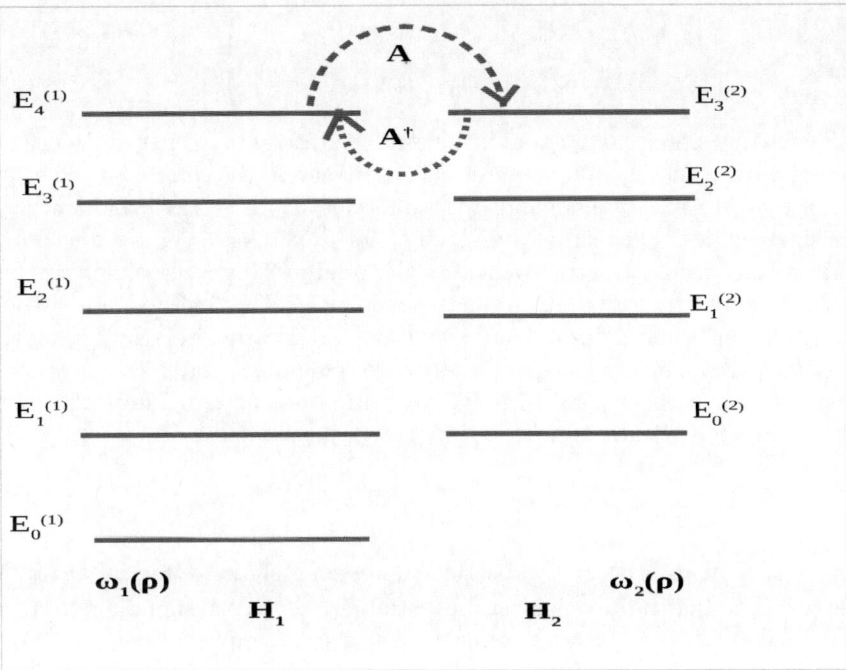

Figure 1. Schematic representation of the energy levels of the supersymmetric partner potentials ω_1 and ω_2 and the action of the operators A and A† on the corresponding eigen functions. The spectrum of ω_1 and ω_2 are identical except that ω_1 has an extra state at zero energy $E_0^{(1)}$=0.

As evident from Eq.(19), H_1 and H_2 have exactly same spectra except for the fact that H_2 has one less bound state (which is the partner state of H_2 corresponding to the ground state of H_1). Hence, from Eqs. (19)-(21) it can be concluded that the Hamiltonians H_1 and H_2 have identical spectra, except that the partner state of H_2 corresponding to the ground state of H_1 is missing in the spectrum of H_2 [26]. **Thus, the supersymmetric partner potentials ω_1 and ω_2 are not strictly isospectral.**

However one can construct a one parameter (say, δ) family of **strictly** isospectral potentials $\hat{\omega_1}(\delta;\rho)$ taking advantage of the nonlinear nature of Riccati equation (Eqs. (12) and (16)) and the fact that for a given $\omega_2(\rho)$; $\omega_1(\rho)$ and the superpotential $W(\rho)$ are not unique. Following [26-28], it can be shown that the most general superpotential satisfying the Riccati equation for $\omega_2(\rho)$ (Eq. (16)) is given by

$$\hat{W}(\rho) = W(\rho) + \frac{{}^-h}{\sqrt{2m}}\frac{d}{d\rho}ln[I_0(\rho) + \delta] \tag{22}$$

where δ is a constant of integration and

$$I_0(\rho) = \int_{\rho'=0}^{\rho} [\psi_0(\rho')]^2 d\rho' \tag{23}$$

Journal of Physics: Conference Series **1354** (2019) 012003 doi:10.1088/1742-6596/1354/1/012003

in which $\psi_0(\rho)$ is the normalized ground state wave function of $\omega_0(\rho)$. Then the potential

$$\left.\begin{aligned}\hat{\omega_1}(\delta;\rho) &= \hat{W}^2(\rho) - \frac{\hbar}{\sqrt{2m}}\hat{W}'(\rho)\\ &= \omega_1(\rho) - 2\frac{\hbar}{2m}\frac{d^2}{d\rho^2}ln[I_0(\rho)+\delta]\end{aligned}\right\} \quad (24)$$

has the same SUSY partner potential $\omega_2(\rho)$ and the ground state at zero energy. Hence $\hat{\omega}_1(\delta;\rho)$ is strictly isospectral with $\omega_1(\rho)$. The parameter δ is arbitrary in the intervals $-\infty < \delta < -1$ and $0 < \delta < \infty$. Since $I_0(\rho)$ lies between 0 and 1, the interval $-1 \leq \delta \leq 0$ is forbidden, for which $\hat{\omega}_1(\delta;\rho)$ will have singularities. One can easily see that for $\delta \to \pm\infty$, $\hat{\omega} \to \omega$ and for $\delta \to 0+$, $\hat{\omega_1}$ develops a narrow and deep attractive well near the origin. This well-barrier combination effectively traps the particle giving rise to a sharp resonance. This method has been tested successfully for a three-dimensional finite square well [29] with parameters such that the well can support one or more resonance state(s) in addition to one bound state.

To calculate the resonance energy accurately, we compute the probability of finding the system within the well region of the potential $\hat{\omega_1}(\delta;\rho)$,

$$P(E) = \int_{\rho'=0}^{\rho_B} |\hat{\psi_E}(\rho',\delta)|^2 d\rho' \quad (25)$$

where ρ_B is the position of the top of the barrier component of the potential $\hat{\omega_1}(\delta;\rho)$ for a choosen δ. Here $\hat{\psi_E}(\rho,\delta)$, the solution for the potential $\hat{\omega_1}(\rho;\delta)$, corresponding to a positive energy E is normalized to have a constant amplitude in the assymptotic region. Plot of the quantity P(E) as a function of E (E > 0) shows a peak at the resonance energy $E = E_R$. Here, choice of δ to be made judiciously in order to avoid numerical errors creeping in the wavefunction in the extremely narrow well for $\delta \to 0+$.

The width (Γ) of the resonance is obtained from the mean life of the state using the energy-time uncertainty relation. The mean life is the reciprocal of the decay constant, which is expressed as a product of the frequency of hit per unit time on the barrier and the probability of tunneling through the barrier. Γ is obtained using the relation

$$\Gamma = 2\sqrt{\frac{\hbar^2}{2\mu}\frac{exp(-2\int_b^c d\rho[\frac{2\mu}{\hbar}(\hat{\omega}(\delta;\rho) - E_R)]}{\int_a^b \frac{d\rho}{\sqrt{(E_R - \hat{\omega}(\delta;\rho))}}}} \quad (26)$$

where a and b are the classical turning points of the nuclear well, while b and c are the classical turning points of the barrier in $\hat{\omega_1}(\delta;\rho)$, corresponding to the resonance energy E_R.

4. Results and Discussions

Ground state of ^{22}C is a $T = 1, J^\pi = 0^+$ state and there exists a resonance state of the same J^π. Thus, the forgoing procedure starting from the ground state of ^{22}C will give $T = 1, J^\pi = 0^+$ resonance(s). Standard supersoft core GPT potential [30] is chosen for the n-n pair while for the core-nucleon (^{20}C-n) pair choose four component SBB [31] Gaussian type potential. The s-wave component of SBB potential is kept repulsive to simulate Pauli principle between the core nucleons and the extra-core neutrons. Here, the parameters of the SBB core-n potential is choosen following our previuos work [23] to reproduce the observed values of the ground state of ^{22}C. The SBB potential has the form

$$V_{Core-n}(\vec{r}) = \sum_l V_{Core-n}^{(l)} \exp(-(r/b_{Core-n}^{(l)})^2) + V^{(ls)}(\vec{l}.\vec{s})\exp(-(r/b_{Core-n})^2) \quad (27)$$

where $V_{Core-n}^{(l)}$ is +50 MeV, -47.32 MeV, -23.0 MeV for l=0, 1 and 2 respectively and $V^{(ls)}$ = -5.855 MeV and $b_{Core-n} = 2.32 fm$ for all the four components. As the original core-n potential

Journal of Physics: Conference Series **1354** (2019) 012003 doi:10.1088/1742-6596/1354/1/012003

gives overbinding we adjusted the range of core-n potential by about 10% to get the resonant 0^+ state of ^{22}C at about 1.27 MeV above the ground (0^+) state. With these potentials, we first solve Eq.(6) for the ground state using HAA. The two-neutron separation energy (B) and the RMS matter radius (R_A) for gradually increasing N_{max} are presented in Coulmns 2 and 4 of Table 1. The results indicate a clear convergence trend in B, but even at $N_{max} = 24$ calculated B is far away from convergence. For this reason, we applied the hyperspherical extrapolation technique [21-23], based on the hyperspherical convergence theorem [20] to get the energy at sufficiently large N_{max}. The extrapolated binding energies with increasing N_{max} is presented in Table 2. Contribution of different partial waves are also presented in columns 5-9 of Table 1 which indicate larger contribution to the ground state energy from the even orbital angular momentum values. The final extrapolated value of energy (=0.4224 MeV) shown in column 3 of Table is in excellent agreement with the experimental value of 0.42 ± 0.94 MeV [8]. The calculated R.M.S. matter radius also agrees fairly with the observations [8, 32-33].

We next utilize the property of the isospectral potential to find the resonant state. The three-body effective potential $\omega_0(\rho)$ when plotted against ρ presents a shallow well followed by a broad and low barrier as shown in Figure 1. This low well-barrier combination can give rise to resonant states. However, since the well is very shallow and the barrier is not sufficiently high, the resonance width is very large and a numerical calculation of the resonant state is quite cumbersome.

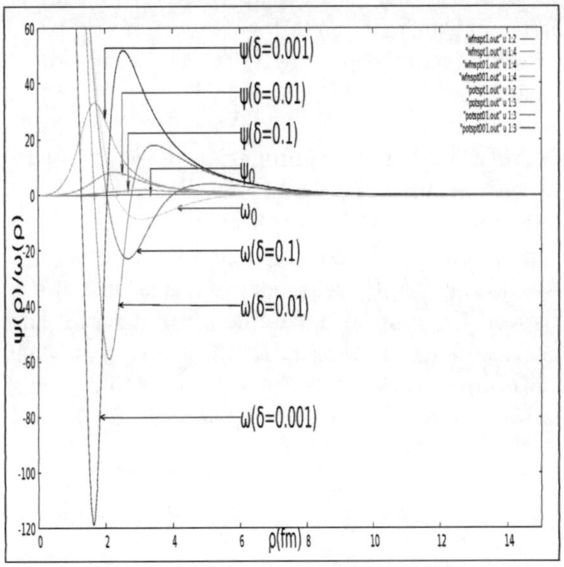

Figure 2. Graphical presentation of isospectral potentials and corresponding wavefunctions for $\delta = \infty, 0.1, 0.01, 0.001$

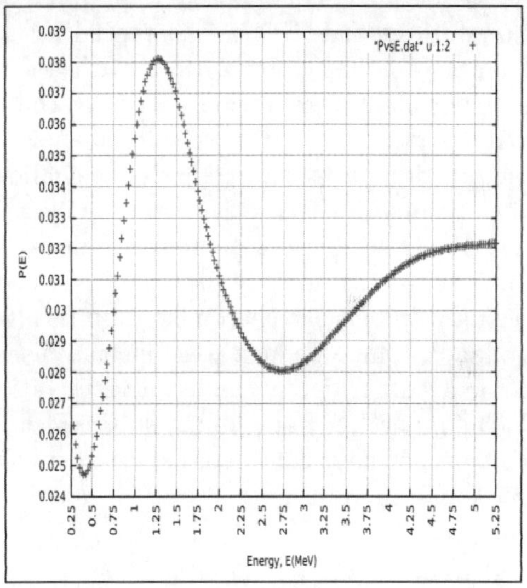

Figure 3. Representative plot of the probality of trapping, P(E) as a function of energy to investigate the resonance energy, E_R for the $J^\pi = 0^+$ state of ^{22}C.

We construct the one-parameter family of isospectral potentials $\hat{\omega}(\delta;\rho)$ according to Eq.(24) by judicious choice of an appropriate value of δ, that gives rise to a narrow and deep well followed by a high barrier. Some of them are demonstrated in Figure 1 as representative cases. This well-barrier combination effectively traps the particles to form a strong resonant state. Calculated parameters of the isospectral potential for a number of δ values, along with the original one $\omega(\rho)$ are presented in columns 2-5 of Table 3. From this, we can see that when δ decreases from 0.1 to 0.001, the depth of the well increases from -22.1 MeV at 2.67 fm to -117.2 MeV at 1.65 fm while the height of the barrier jumps from 4.2MeV at 5.1fm to 51.5 MeV at 2.56 fm. This

Journal of Physics: Conference Series **1354** (2019) 012003 doi:10.1088/1742-6596/1354/1/012003

demonstrates the dramatic effect on the lowest eigen potential $\hat{\omega}_1(\delta;\rho)$ as δ approaches 0+. Still smaller positive values of δ are not desirable since the well will be too narrow to calculate the wave function accurately by a numerical procedure.

We calculated the probability of P(E) of the system to be trapped within the well-barrier combination for a range of energies E. A representative plot of P(E) versus E shown in Figure 3, exhibits resonance peak at the energy $E_R \simeq 1.27$ MeV.

The calculated widths of resonance for different $\delta \to 0+$ indicates dramatic enhancement in trapping effect of $\hat{\omega}_0(\delta;\rho)$ for smaller values of delta. From column 6 of Table 3, it can be seen that the width of the resonance decreases significantly with decrease in δ values. Hence for the original potential $\omega_0(\rho)$ (which corrrsponds to $\delta \to +\infty$), the exact value of resonance energy becomes uncertain. But the uncertainty in E_R decreases as δ decreases towards 0+. The enhancement of accuracy in determination of E_R is the principal advantage of using Supersymmetric formalism. Since $\hat{\omega}_1(\delta;\rho)$ is strictly isospectral with $\omega_1(\rho)$, any value of δ is admissible in principle. However, a judicious choice of δ is necessary for accurate determination of the resonance energy. Results of the present calculations are summarized in Table 4 and compared with other works found in literature.

5. Summary and Conclusions

In this work have employed hyperspherical harmonics expansion method to investigate the bound state properties of the loosely bound nucleus ^{22}C in the framework of three-body (^{20}C+n+n) system. Standard GPT [30] potential is used for the $n-n$ pair while for the remaining pairs we employed SBB potential [31] with a slightly modifide range parameter so as to reproduce the experimental excitation energy of the 0^+ state. The lowest-eigen potential $\omega_0(\rho)$ for the ground state ($J^\pi = 0^+$) exhibits shallow well followed by a skinny wide barrier (Figure 2). Computation of resonances in such a shallow well and wide barrier combination presents a very difficult numerical task. However, in our scheme we managed to circumvent this difficulty by using the isospectral potential technique by judicious choice of the free parameter δ, which gives rise to a sufficiently deep and narrow well surrounded by relatively high and thin barrier. This narrow well-barrier combination facilitates trapping effect thereby resulting in a prominent resonance. Thus, we may conclude that use of isospectral potential technique is an essential one for the accurate determination determination of resonanances in weakly bound exotic systetms which support only one bound state. Even, the technique can can also be extended to systems which do not have any bound state by invoking suitable modification in it so as to fit for bound state in continuum.

6. Tables

Table 1. Calculated energy (B=-E) with corresponding relative convergence trend ($\chi = \frac{B(N_{max}+4)-B(N_{max})}{B(N_{max}+4)}$), rms matter radius ($R_A$), and partial contribution of different l_x's on energy with increasing N_{max} for the ground ($J^\pi = 0^+$) state of ^{22}C.

N_{max}	Energy, -E (MeV)	χ	R_A (fm)	$-E_{l_\xi}(MeV)$				
				$l_\xi = 0$	$l_\xi = 1$	$l_\xi = 2$	$l_\xi = 3$	$l_\xi = 4$
12	0.2419	0.2032	3.4259	0.21449	0.00678	0.02309	0.00016	0.00211
16	0.3035	0.1185	3.4217	0.27247	0.00753	0.02469	0.00018	0.00231
20	0.3443	0.0719	3.4187	0.31169	0.00819	0.02536	0.00020	0.00237
24	0.3710		3.4163	0.33735	0.00875	0.02550	0.00022	0.00241

Table 2. Extrapolated energy ($B_{Extrapolated}$) and the corresponding relative convergence trend (χ) with increasing N_{max} for the ground ($J^{\pi} = 0^{+}$) state of ^{22}C.

N_{max}	$B_{Extrapolated}$	χ	N_{max}	$B_{Extrapolated}$	χ
24	0.3710825	0.072043208	96	0.4224221	0.000033559
28	0.3886654	0.045239185	100	0.4224314	0.000022071
32	0.4002297	0.028894056	104	0.4224375	0.000014516
36	0.4078355	0.018649189	108	0.4224416	0.000009547
40	0.4128378	0.012116931	112	0.4224442	0.000006279
44	0.4161279	0.007906285	116	0.4224459	0.000004129
48	0.4182917	0.005173055	120	0.4224471	0.000002716
52	0.4197149	0.003390777	124	0.4224479	0.000001786
56	0.4206509	0.002225148	128	0.4224484	0.000001175
60	0.4212665	0.001461339	132	0.4224487	0.000000773
64	0.4216714	0.000960198	136	0.4224489	0.000000508
68	0.4219377	0.000631123	140	0.4224491	0.000000334
72	0.4221128	0.000414916	144	0.42244920	0.000000219
76	0.4222280	0.000272816	148	0.42244921	0.000000145
80	0.4223038	0.000179398	152	0.42244925	0.000000095
84	0.4223536	0.000117976	156	0.42244927	0.000000063
88	0.4223864	0.000077587	160	0.42244929	0.000000041
92	0.4224079	0.000051026	164	0.42244931	0.000000027

Table 3. Parameters of the isospectral potential, $\hat{\omega}_0(\delta;\rho)$ as the parameter δ decreases from $+\infty$ (for the original potential $\omega_0(\rho)$) towards 0+.

δ	Potential well		Potential Barrier		Resonance width (Γ)
	Depth(MeV)	Position(fm)	Height(MeV)	Position(fm)	
1000000	-8.298	3.068	0.623	6.579	
100	-8.321	3.068	0.627	6.576	
50	-8.343	3.067	0.630	6.573	
1	-10.404	3.008	1.041	6.372	18937344.7486
0.1	-22.098	2.674	4.181	5.101	6.242
0.001	-117.207	1.648	51.541	2.555	0.134
0.0001	-185.743	1.211	103.451	1.993	1.543E-03
0.00001	-337.228	0.879	162.829	1.456	2.504E-06
0.000001	-596.238	0.632	300.474	1.076	3.788E-10

Table 4. Comparison of the calculated results with experimental data and other reference works found in the literature for ^{22}C.

State	Observables	Present work	Others work	Reference
0^+	E	-0.422 MeV	-0.420 ±0.940 MeV	[8,18]
	R_A	3.42 fm	5.4±0.9 fm	[8]
			3.03 – 3.34 fm	[32]
			3.58 – 3.74 fm	[33]
0_1^+	E_R	1.27 MeV	1.02 MeV	[18]
	Γ	0.51 MeV	0.52 MeV	[18]

Acknowledgments

The authors acknowledge the computational facilities extended by Aliah University, India.

References

[1] I. Tanihata et al 1985 Phys. Rev. Lett. **55** N0.24, 2676
[2] N. Kobayashi *et al* 2012 *Phys. Rev. C.* **86** 054604
[3] W. S. Hwash 2017 *Turk. J. Phys.* **41** 151-159
[4] A. S. Jensen and K. Riisager 2000 Phys. Lett. **B480**, No. 1, 39-44
[5] W. Schwab et al 2000 Z. Phys. A350 (1995) 283
[6] G. O. Audi et al 2003 NPA No.1,3-128
[7] B. Acharya, C. Ji, D.R. Phillips 2013 *Phys. Lett. B.* **723** 196
[8] K. Tanaka *et al.* 2010 *Phys. Rev. Lett.* **104** 062701
[9] Daniel Sääf and Christian Forssen (2014) Phys. Rev. **C89**, No.1, 011303
[10] A. V. Nesterov et al 2010 *Phys. of Part. and Nuclei* **41** No.5, 716-765
[11] S. Korennov and Pierre Descouvemont 2004 Nucl. Phys. A740, No.3, 249-267
[12] A. Cobis, D.V. Federov and A.S. Jensen 1997 Phys. Rev. Lett. **79**, 2411
A. Cobis, D.V. Federov and A.S. Jensen 1998 Phys. Rev. C **58**, 1403
[13] A. Csótó 1993 Phys. Lett. **B315**. 24; 1993 Phys. Rev. **C48** 165; A. Csótó 1993 *ibid.* **C48** 165; A. Csótó 1994 *ibid.* **C49** 3035
[14] S. Aoyoma, S. Mukai, K. Kato and K. Ikeda 1995 Prog. Theor. Phys. **94** 343
[15] Danilin, B. V., Rogde, T., Ershov, S. N., Heiberg-Andersen, H., Vaagen, J. S., Thompson, I. J.,Zhukov, M. V. 1997 Phys. Rev. C55, R577
[16] N.Tanaka, Y. Suzuki, K. Varga 1997 Phys. Rev. C56, 562
[17] V. Vasilevsky, A. V. Nesterov, F. Arickx, J.Broeckhove 2001 Phys. Rev. C63, 034607
[18] Kazuyuki Ogata, Takayuki Myo, Takenori Furumoto, Takuma Matsumoto, and Masanobu Yahiro 2013 *Phys. Rev. C.* **88** 024616
[19] Ballot and M. Fabre de la Ripelle 1982 Ann. Phys. **127**, 62 ;
T.K. Das, H.T. Coelho and M. Fabre de la Ripelle 1982 Phys. Rev. C **26**, 2288
[20] T.R. Schneider 1972 Phys. Letts. B **40**, 439
[21] T.K. Das, R. Chattopadhyay and P.K. Mukherjee 1994 Phys. Rev. A **50**, 3521
[22] Md. A. Khan 2012 *Eur. Phys. Jour.D* **66** 83
[23] Md. A. Khan and T.K. Das 2001 Pramana- J. Phys. **57**, 701
[24] J.L. Ballot, M. Fabre de la Ripelle and J.S. Levinger 1982 Phys. Rev. C **26**, 2301
[25] T.K. Das, H.T. Coelho and M. Fabre de la Ripelle 1982 Phys Rev. C **26**, 2281
[26] F. Cooper, A. Khare and U. Sukhatame 1995 Phys. Rep. **251**, 267
[27] A. Khare, U. Sukhatme 1989 J. Phys. A **22**, 2847
[28] M. M. Nieto 1984 Phys. Lett. **145**, 208
[29] T.K. Das and B.Chakrabarti 2001 Phys. Letts. A **288**, 4
[30] D. Gogny, P. Pires and R. de Tourreil 1970 Phys. Letts. **32B**, 591
[31] S. Sack, L.C. Biedenham and G. Breit 1954 Phys. Rev. **93**, 321
[32] T. Inakura, W. Horiuchi, Y. Suzuki and T. Nakatsukasa 2014 Phys. Rev. C **89**, 064316
[33] W. Horiuchi and Y. Suzuki 2006 Phys. Rev. C **74**, 034311

Asymptotically-flat Black Hole in EiBI-Born-Infeld Theory

Byon N. Jayawiguna, Handhika S. Ramadhan
Department of Physics, FMIPA, University of Indonesia, Kampus UI Depok, Indonesia
E-mail: byon.nugraha@ui.ac.id, hramad@ui.ac.id

Abstract. We study static spherically-symmetric solutions of charged Eddington-inspired Born-Infeld (EiBI) theory of gravity in asymptotically-flat D-dimensional spacetime. We consider nonlinear electrodynamics for the matter fields. In this particular work, the nonlinear theory we specifically consider is the Born-Infeld electrodynamics. We found that metrics can not be integrated in general dimensions. In order to solve it, we focus on studying on the case $\alpha \equiv 4\kappa b^2 = 1$. We study the electric fields. We find that this theory produces finite electric fields for any dimension. In Ricci scalar calculations, we find only point singularity.

1. Introduction

Black hole is an interesting object in the field of gravity and cosmology. The concept of this black hole appears in Einstein's field equation solution. When we consider electromagnetic field as a source of matter sector to construct black hole solution in Einstein field equations, one usually use Maxwell electrodynamics. However, if we look at a very strong field, Maxwell's theory does not apply. In 1930, Born-Infeld [1] in his proposal on electrodynamics found a non-linear form with the aim of obtaining a finite value on the electron's self-energy. This makes scientists interested in researching more about this topic. Many proposals on Born-Infeld is widely applied to black hole object. Demianski finds a solution from Einstein-Born-infeld in 4 dimensions in asymptotically-flat space-time and finds that when the internal mass is zero, the solution is regular and describes static electromagnetic geon [2]. This led many researchers to further analyze with the addition of cosmological constant [3] or the addition of dimensions [4, 5].

In 2010, Eddington-inspired Born-Infeld (EiBI) is an alternative theory of gravity proposed by Banados and Ferreira [6]. This alternative theory successfully covered the singularity in the early universe. The characteristic of EiBI theory is in vacuum ($\kappa \rightarrow 0$) , the action of EiBI theory will be reduced to Einstein-Hilbert action. When discussing a non-vacuum case ($T_{\mu\nu} \neq 0$), there will be a deviation from the general theory of relativity. Although still fairly new, this EiBI theory has been widely studied in black hole and other compact stars both 4d or higher dimension [7, 8, 9, 10, 11].

In this paper, we will study higher dimensional EiBI theory with Born-infeld as a matter sector. The $4d$ has been studied by Jana-Kar [12]. In this work, we restrict ourselves to finding the exact solution in the electrostatic ansatz with zero cosmological constant Λ. This paper is structured as follows. Section 2 discusses EiBI Formalism. Section 3 discusses the structure of

Content from this work may be used under the terms of the Creative Commons Attribution 3.0 licence. Any further distribution of this work must maintain attribution to the author(s) and the title of the work, journal citation and DOI.
Published under licence by IOP Publishing Ltd

higher dimensional EiBI-Born-Infeld. In section 4 we will discuss metric solution, electric fields and the curvature in each dimension. Finally, some concluding remarks are given in Section 5

2. EiBI Action

We start with EiBI action [6] where in vacuum limit, this action reduces to Einstein-Hilbert action,

$$S(g,\Gamma,\Phi) = \frac{1}{8\pi\kappa}\int d^Dx\left(\sqrt{-|g_{\mu\nu}+\kappa R_{\mu\nu}(\Gamma)|} - \lambda\sqrt{-|g_{\mu\nu}|}\right) + S_M(g,\Phi). \quad (1)$$

Where $\lambda = 1+\kappa\Lambda$, $q_{\mu\nu}$ is auxiliary metric and κ is so called EiBI parameter. Here, we set $G=1$. From equation (1), varying the action with respect to pure metric g can appear a ghost term. In order to avoid it, Vollick [13] use Palatini formalism and he treat Γ and g as two independent fields and Ricci tensor is depent on Γ,

$$q_{\mu\nu} = g_{\mu\nu} + \kappa R_{\mu\nu}(\Gamma). \quad (2)$$

The fields equation from EiBI theory can be obtained from varying EiBI action with respect Γ and leads us to $\nabla_\beta q_{\mu\nu} = 0$. After some algebra, we get

$$\Gamma^\sigma_{\alpha\beta} = \frac{1}{2}q^{\rho\sigma}\left(\partial_\alpha q_{\beta\rho} + \partial_\beta q_{\rho\alpha} - \partial_\rho q_{\alpha\beta}\right). \quad (3)$$

If we varying the action with respect to metric tensor $g_{\mu\nu}$, we get EiBI equation

$$\sqrt{-q}q^{\mu\nu} = \lambda\sqrt{-g}g^{\mu\nu} - 8\pi\kappa\sqrt{-g}T^{\mu\nu}. \quad (4)$$

with q and g is determinant of $q_{\mu\nu}$ and $g_{\mu\nu}$, $T^{\mu\nu}$ is energy-momentum tensor and κ is EiBI parameter.

3. Structure of Higher Dimensional Charged Black Hole in EiBI-Born-Infeld Theory

We start with Born-Infeld lagrangian for matter fields,

$$\mathcal{L}_{BI} = \frac{b^2}{4\pi}\left[1 - \sqrt{1+\frac{F}{b^2}}\right], \quad (5)$$

with $F = \frac{1}{2}F_{\mu\nu}F^{\mu\nu}$ and b is Born-Infeld parameter. When $b\to\infty$, Maxwell's theory was recovered with higher-order corrections,

$$\mathcal{L}_M = -F_{\mu\nu}F^{\mu\nu} + O(F^4). \quad (6)$$

Hence, the resulting of tensor energy-momentum is

$$T_{\mu\nu} = -\frac{b^2}{4\pi}\left[g_{\mu\nu}\left(\sqrt{1+\frac{F_{\alpha\beta}F^{\alpha\beta}}{2b^2}} - 1\right) - \frac{F_{\mu\sigma}F^\sigma_\nu}{b^2\sqrt{1+\frac{F_{\alpha\beta}F^{\alpha\beta}}{2b^2}}}\right]. \quad (7)$$

In this paper, we limit our work in electrostatic $A_\mu = \{\phi,0,0,0\}$. Varying equation (1) with respect to A_μ yields

$$\nabla_\mu\left(\frac{F^{\mu\nu}}{\sqrt{1+\frac{F_{\alpha\beta}F^{\alpha\beta}}{2b^2}}}\right) = 0. \quad (8)$$

For spherically symmetric work, we use ansatz

$$g_{\mu\nu}dx^\mu dx^\nu = -U(\bar{r})e^{2\psi(\bar{r})}dt^2 + U(\bar{r})e^{2\nu(\bar{r})}d\bar{r}^2 + V(\bar{r})\bar{r}^2 d\Omega^2_{D-2}, \tag{9}$$

$$q_{\mu\nu}dx^\mu dx^\nu = -e^{2\psi(\bar{r})}dt^2 + e^{2\nu(\bar{r})}d\bar{r}^2 + \bar{r}^2 d\Omega^2_{D-2}, \tag{10}$$

where $d\Omega^2_{D-2}$ denotes the metric of an unit $(D-2)$ sphere. From ansatz above, $\bar{r}$ is radial coordinate, but the physical radial distance is $r^2 = V(\bar{r})\bar{r}^2$. This physical radial distance is interpreted as area of $(D-2)$ sphere. Here, this solution can't solved easily, so we follow Jana-Kar [12] ansatz. The advantage of this metric is absence of differential expression in field equation (4). Hence, as we shall see, $U(\bar{r})$ and $V(\bar{r})$ can be solved in polynomial equation. Before evaluate these field equation, its important to solving the electromagnetic equation first. In electrostatics, the non-zero component of Maxwell stress tensor $F_{\mu\nu}$ is $F_{10} = \partial_r\phi$. Solving the electromagnetic equation, we obtain

$$E(\bar{r}) = -\frac{d\phi}{d\bar{r}} = \frac{qUe^{\nu+\psi}}{\sqrt{V^{D-2}\bar{r}^{2(D-2)} + \frac{q^2}{b^2}}}, \tag{11}$$

with q an integration constant which can usually be identified as the charge. Hence, the non-zero of tensor energy-momentum is

$$T^{00} = \frac{e^{-2\psi}b^2}{U4\pi}\left(\frac{\sqrt{V^{D-2}\bar{r}^{2(D-2)} + \frac{q^2}{b^2}}}{V^{\frac{D-2}{2}}\bar{r}^{D-2}} - 1\right), \tag{12}$$

$$T^{11} = -\frac{e^{-2\nu}b^2}{U4\pi}\left(\frac{\sqrt{V^{D-2}\bar{r}^{2(D-2)} + \frac{q^2}{b^2}}}{V^{\frac{D-2}{2}}\bar{r}^{D-2}} - 1\right), \tag{13}$$

$$T^{22} = \frac{b^2}{4\pi V\bar{r}^2}\left(1 - \frac{V^{\frac{D-2}{2}}\bar{r}^{D-2}}{\sqrt{V^{D-2}\bar{r}^{2(D-2)} + \frac{q^2}{b^2}}}\right), \tag{14}$$

$$T^{33} = \frac{T^{22}}{\sin^2\theta}, \qquad T^{ab} = \frac{T^{22}}{\prod_{j=1}^{D-2}\sin^2\theta_j}. \tag{15}$$

Inserting this tensor energy-momentum to equation (4) and after some algebra, we get the expression of $V(\bar{r})$ and $U(\bar{r})$ from polynomial equation

$$(1-\alpha)V^{D-2} - (2-\alpha)V^{\frac{D-2}{2}} + 1 - \frac{\kappa q^2\alpha}{\bar{r}^{2(D-2)}} = 0. \tag{16}$$

Where $\alpha \equiv 4\kappa b^2$. Of course, there are many possible solution for $V(\bar{r})$ and also $U(\bar{r})$. We can manipulate this solution into two possible solution. Therefore, we have

$$V(\bar{r}) = \left[\frac{1 - \frac{\alpha}{2}\left(1 + \sqrt{1 + \frac{q^2}{b^2\bar{r}^{2(D-2)}} - \frac{4\kappa b^2}{\bar{r}^{2(D-2)}}}\right)}{(1-\alpha)}\right]^{\frac{2}{D-2}}, \tag{17}$$

$$U(\bar{r}) = \frac{\left(1-\frac{\alpha}{2}\right)\sqrt{1 + \frac{4\kappa q^2}{\alpha\bar{r}^{2(D-2)}} - \frac{4\kappa q^2}{\bar{r}^{2(D-2)}}} - \frac{\alpha}{2}}{(1-\alpha)\sqrt{1 + \frac{4\kappa q^2}{\alpha\bar{r}^{2(D-2)}} - \frac{4\kappa q^2}{\bar{r}^{2(D-2)}}}}V(\bar{r})^{\frac{4-D}{2}}. \tag{18}$$

On the other hand, from (2) one obtains

$$\left(\frac{1-U}{\kappa}\right)e^{2\psi} = e^{-2\nu+2\psi}\left(-\frac{(D-2)\psi'}{\bar{r}} + \nu'\psi' - \psi'^2 - \psi''\right), \tag{19}$$

$$\left(\frac{1-U}{\kappa}\right)e^{2\nu} = \left(\frac{(D-2)\nu'}{\bar{r}} + \nu'\psi' - \psi'^2 - \psi''\right), \tag{20}$$

$$\left(\frac{1-V}{\kappa}\right)\bar{r}^2 = (D-3) - \frac{1}{(D-2)\bar{r}^{D-4}}\left(\left(\bar{r}^{D-2}\right)' e^{2\psi}\right)'. \tag{21}$$

The first two equations are solved by $\psi = -\nu$. Inserting it into (21) yields

$$e^{2\psi(\bar{r})} = 1 - \frac{2M}{\bar{r}^{(D-3)}} - \frac{\bar{r}^2}{(D-1)\kappa} + \frac{1}{\kappa\bar{r}^{(D-3)}}\int V(\bar{r})\bar{r}^{(D-2)}d\bar{r}. \tag{22}$$

Finally, equation (22) is the last metric solution that we want to solve. Unfortunately, this equation cannot be solved analitically for arbitrary D. We therefore specify the value of α. In this work, we specify $\alpha = 1$ from polynomial equation (19). Hence, the metric functions V, U and $e^{2\psi}$ becomes

$$V(\bar{r}) = \left(1 - \frac{\kappa q^2}{\bar{r}^{2(D-2)}}\right)^{\frac{2}{D-2}}, \tag{23}$$

$$U(\bar{r}) = \left(1 + \frac{\kappa q^2}{\bar{r}^{2(D-2)}}\right)V(\bar{r})^{\frac{4-D}{2}} \tag{24}$$

and

$$e^{2\psi}(\bar{r}) = 1 - \frac{2M}{\bar{r}^{(D-3)}} - \frac{\bar{r}^2}{(D-1)\kappa} + \frac{\bar{r}^2\,{}_2F_1\left(-\frac{2}{D-2}, -\frac{D-1}{2(D-2)}; \frac{D-3}{2(D-2)}; q^2\bar{r}^{4-2D}\kappa\right)}{(D-1)\kappa}. \tag{25}$$

Where ${}_2F_1(a, b; c; d)$ is hypergeometric function. When $q \to 0$, the solution reduces to the standard Tangherlini form. Equations (23)-(25) are the exact solution of higher dimensional Born-Infed gravity coupled to Born-Infeld electrodynamics in radial coordinate $\bar{r}$ at specific value of α. This exact solution comes from gauge (9) that has $\bar{r}$ on its metric. Fortunately, it is possible to bring it into another gauge that has r on metric functions, called Tangherlini gauge. Since the physical radial distance is $r^2 = \bar{r}^2 V(\bar{r})$, we therefore set $r^2 = \left(1 - \frac{\kappa q^2}{\bar{r}^{2(D-2)}}\right)^{\frac{2}{D-2}}\bar{r}^2$. The resulting metric is

$$ds^2 = -A(r)f(r)dt^2 + \frac{r^{D-4}\left(\frac{1}{2}\right)^{\frac{2}{D-2}}A(r)\left(1+\sqrt{B(r)}\right)^{\frac{2}{D-2}}}{B(r)f(r)}dr^2 + r^2 d\Omega^2, \tag{26}$$

where

$$A(r) = \left(1 + \frac{\kappa q^2}{y(r)^{2(D-2)}}\right)\left(1 - \frac{\kappa q^2}{y(r)^{2(D-2)}}\right)^{\frac{4-D}{D-2}}, \tag{27}$$

$$f(r) = 1 - \frac{2M}{y(r)^{(D-3)}} - \frac{y(r)^2}{(D-1)\kappa} + \frac{y(r)^2\,{}_2F_1\left(-\frac{2}{D-2}, -\frac{D-1}{2(D-2)}; \frac{D-3}{2(D-2)}; \frac{\kappa q^2}{y(r)^{2(D-2)}}\right)}{(D-1)\kappa}, \tag{28}$$

$$y(r) \equiv \bar{r}(r) = \left(\frac{1}{2}\right)^{\frac{1}{D-2}}\left(1+\sqrt{B(r)}\right)^{\frac{1}{D-2}} r, \tag{29}$$

and

$$B(r) = 1 + \frac{4\kappa q^2}{r^{2(D-2)}}. \tag{30}$$

Equation (26)-(30) are exact solutions in D-dimension which we want to analyze its behaviour. On the other hand, a curvature of this space-time is important to be analyzed. Also, electric fields that we have shown in (11) has interesting feature. All of these quantities will be discuss deeply in the next section.

4. Result and Discussion

From the previous section, we study static spherically-symmetric solutions of EiBI-Born-Infeld in higher dimension. Equation (26)-(30) constitute an exact solution in this paper. Since black holes are containing singularity that enclosed by horizon, it is interesting to get the information about the radius of its horizon by study the metric solution. For $D \geq 4$, metric solution in Einstein-Born-Infeld has been studied in [5] and found that these metric behaviour causing double horizons even in asymptotically-flat space-time. While in EiBI (Eddington inspired Born-Infeld) is absent with different values of parameter κ. Next, we will present the metric behaviour of EiBI-Born-Infeld black hole in $D \geq 4$.

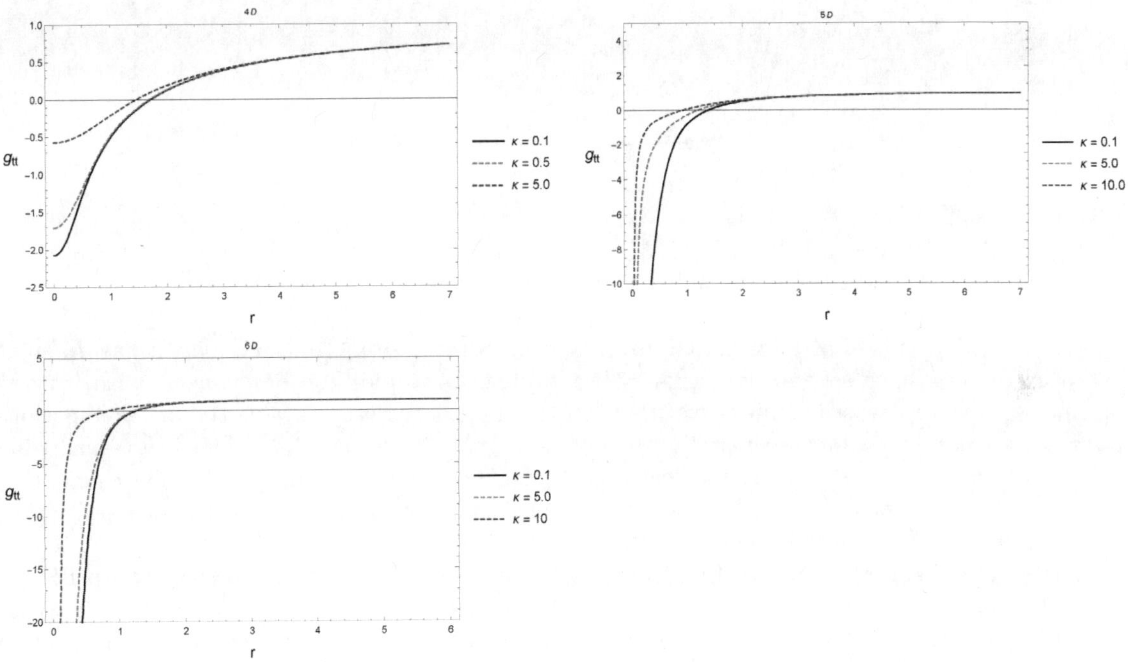

Figure 1. Metric functions of g_{tt} as a function of r with $M = 1$ and $q = 0.7$ in asymptotically-flat space-time.

In Figure 1, we show the metric of $g_{tt} = U(r)e^{2\psi(r}$ in specify dimension. Important to note that this plot as a function of r since a physical radius distance is $r = \bar{r}\sqrt{V(\bar{r})}$, as we mention before. We varying the value of κ from weak to strong coupling. Based on this plot, $4d$ metric shape is the same as Jana-kar in $\alpha = 1$. It turns out that the behaviour of this metric in $D > 4$ is also the same, 1 horizon for different κ. We note a generic feature that for κ positive the horizon radius r decreases with increasing κ. In $5d$ and $6d$, also have the same amount of horizon and size on its radius (decreases with increasing κ). Note that the $D = 4$ solution we found

Journal of Physics: Conference Series **1354** (2019) 012004 doi:10.1088/1742-6596/1354/1/012004

looks regular at the origin [12]. The regularity of this metric seems to indicate that the EM source is point-charge. In higher dimensions ($D \geq 5$), all the metric solutions are singular at the origin. It is implied that the metric functions are limited to minimum radius where the charged is distributed. So there is no discontinuity between $D = 4$ and $D \geq 5$ as $r \to 0$. To ensure this statement, as can we see later, the scalar Ricci is singular at the origin in all dimensions. Thus, we interpreted our solutions as a genuine black holes. In large r, these metric can be seen as asymptotically-flat space-time.

In our discussion about metric behaviour of asymptotically-flat EiBI-Born-Infeld black hole in $D \geq 4$. To ensure this metric behaviour we use scalar quantities, Ricci scalar $\mathcal{R} = g^{\mu\nu} R_{\mu\nu}$ in this paper. While there are several order that exist in scalar quantitites such as $R_{\mu\nu\rho\sigma} R^{\mu\nu\rho\sigma}$ and $R_{\mu\nu} R^{\mu\nu}$, one can verify that the Ricci scalar is the lowest order. In Figure 2, we plot the Ricci

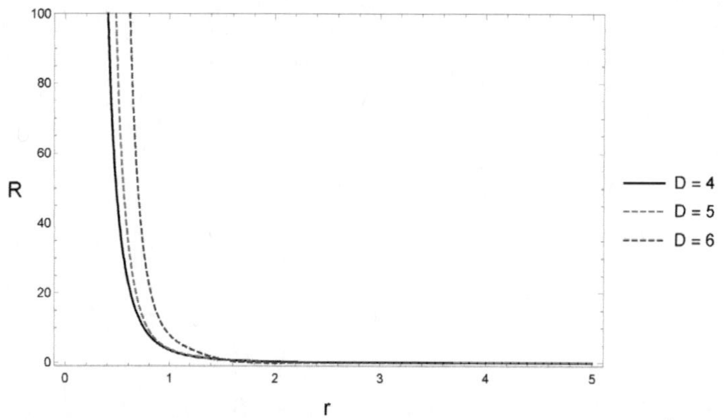

Figure 2. A typical plot of $\mathcal{R} = g^{\mu\nu} R_{\mu\nu}$ with $M = 1$, $q = 1$ and $\kappa = 10$ in asymptotically-flat space-time. The Ricci scalar still suffers singularity at the origin.

scalar as a function of physical radial distance r. According to metric functions, it has different feature in $r \to 0$ in several dimensions. It motivate us to plot the Ricci scalar with varying dimension to test whether this curvature has the same characteristics with the metric functions or not. This typical plot, fortunately, still suffers singularity at the origin. This is not surprising since black hole has singularity that enclosed by horizon. However, this Ricci scalar calculation does not ensure the singularity of the electric field. In Reissner-Nordstrom, one can verify that the strength of the electric fields blow up near origin. This problem is covered by studying Born-Infeld electrodynamics in Einstein gravity, based on Born-infeld proposal. Surprisingly, EiBI come up with its remarkable feature when coupled to $4d$ Maxwell electrodynamics that has finite value at the origin [7] and $4d$ Born-Infeld electrodynamics has the same feature with its Einstein gravity, produces finite electric fields in a strong regime [12]. Hence, it is tempting to check whether EM field can still have regular on its self-energy in higher dimension [11]. In Figure 3, we present the solution of electric field in non-linear electrodynamics $E(r)$ in equation (11). Note that when $b \to \infty$ this formalism reduces to electric field in Maxwell electrodynamics. In this plot, the electric fields is a function of physical radial distance r and this solution obtained with $\alpha \equiv 4\kappa b^2 = 1$. Hence, we have a relation between κ and b

$$b^2 = \frac{1}{4\kappa}, \tag{31}$$

where κ is EiBI parameter and b is non-linear parameter. Plots that we present based on the strength of parameter κ and b. The strength is weak when $\kappa \to 0$ and $b > 1$ and strong when $\kappa \geq 1$ and $b < 1$. As we can see, we can't find the electric field that has strong on κ and weak

Journal of Physics: Conference Series **1354** (2019) 012004 doi:10.1088/1742-6596/1354/1/012004

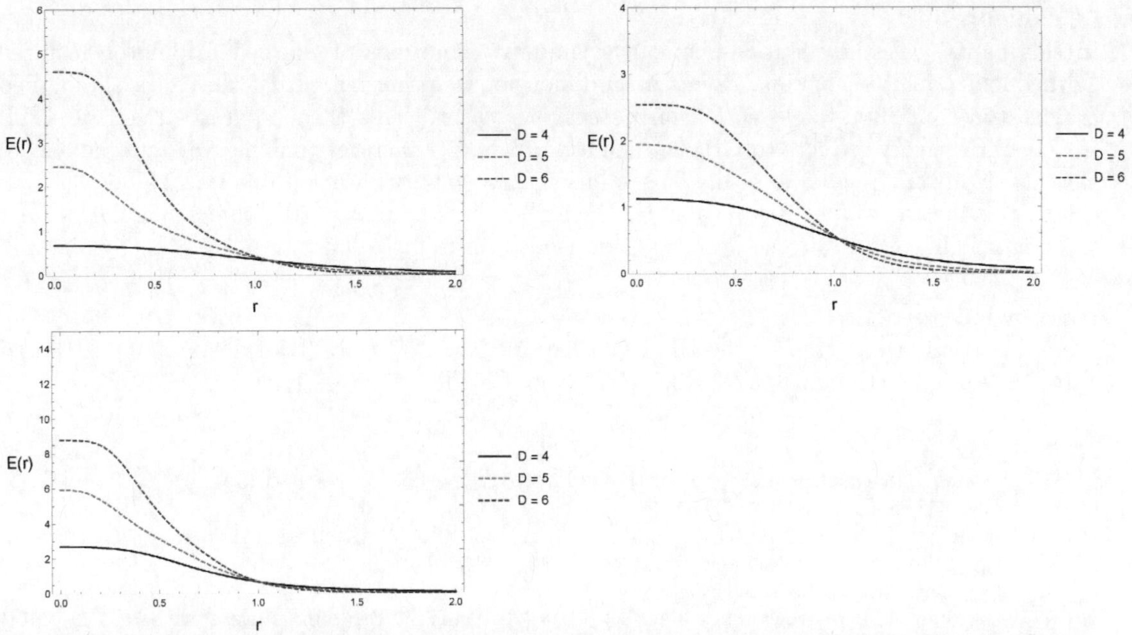

Figure 3. A plot of Born-Infeld electric fields $E(r)$ as a function of r with $q = 0.7$ in asymptotically-flat space-time. This plot is the equation given in (11). For parameter values from left panel is $\kappa = 2$ and $b = 0.35$, right panel is $\kappa = 0.4$ and $b = 0.7$ and left panel at the bottom is $\kappa = 0.1$ and $b = 1.5$.

on b because of its natural fraction on $b^2 = \frac{1}{4\kappa}$. On the other hand, the electric fields in this plot is always has finite value on its self-energy. In large r, $E(r)$ becomes zero.

5. Conclusion

In this paper, we study static spherically-symmetric solutions of charged Eddington-inspired Born-Infeld (EiBI) theory of gravity in D-dimensional spacetime. EiBI theory is a modified gravity that put born-infeld action on its gravity sector. This theory parametrized by κ. If $\kappa \to 0$, EiBI theory reduces to GR. Our solution in $D = 4$, reduces to Jana-Kar and reduces to Reissner-Nordstrom when $\kappa \to 0$. This solution have 1 horizon in all different value of κ. To ensure this metric, we use Ricci scalar $\mathcal{R}$. We found that the metric still suffers singularity near the origin. The electric fields of this solution is also finite in $D > 4$.

Acknowledgment

We are partially supported by the UIs PITTA grant No. 2267/UN2.R3.1/HKP.05.00/2018 and Hibah Q1Q2 UI (NKB-0270/UN2.R3.1/HKP.05.00/2019).

References

[1] M. Born and L. Infeld, Proc. Roy. Soc. Lond. A **144** (1934) 425. doi:10.1098/rspa.1934.0059
[2] M. Demianski, Found. Phys. **16** (1986) 187.
[3] S. Fernando and D. Krug, Gen. Rel. Grav. **35** (2003) 129 doi:10.1023/A:1021315214180 [hep-th/0306120].
[4] R. G. Cai, D. W. Pang and A. Wang, Phys. Rev. D **70** (2004) 124034
[5] T. K. Dey, Phys. Lett. B **595** (2004) 484
[6] M. Banados and P. G. Ferreira, Phys. Rev. Lett. **105** (2010) 011101 Erratum: [Phys. Rev. Lett. **113** (2014) no.11, 119901] [arXiv:1006.1769 [astro-ph.CO]].
[7] H. Sotani and U. Miyamoto, Phys. Rev. D **90** (2014) 124087 [arXiv:1412.4173 [gr-qc]].
[8] S. W. Wei, K. Yang and Y. X. Liu, Eur. Phys. J. C **75** (2015) 253 Erratum: [Eur. Phys. J. C **75** (2015) 331] doi:10.1140/epjc/s10052-015-3556-9, 10.1140/epjc/s10052-015-3469-7 [arXiv:1405.2178 [gr-qc]].
[9] I. Prasetyo, I. Husin, A. I. Qauli, H. S. Ramadhan and A. Sulaksono, JCAP **1801** (2018) no.01, 027 doi:10.1088/1475-7516/2018/01/027
[10] A. I. Qauli, M. Iqbal, A. Sulaksono and H. S. Ramadhan, Phys. Rev. D **93** (2016) no.10, 104056 doi:10.1103/PhysRevD.93.104056
[11] B. N. Jayawiguna and H. S. Ramadhan, Nucl. Phys. B **943** (2019) 114615 doi:10.1016/j.nuclphysb.2019.114615
[12] S. Jana and S. Kar, Phys. Rev. D **92** (2015) 084004 [arXiv:1504.05842 [gr-qc]].
[13] D. N. Vollick, Phys. Rev. D **69** (2004) 064030

Journal of Physics: Conference Series **1354** (2019) 012005 doi:10.1088/1742-6596/1354/1/012005

Creation of the de sitter space from nothing in Eddington-inspired Born-Infeld theory

Reyhan D. Lambaga, Handhika S. Ramadhan
Department of Physics, FMIPA Universitas Indonesia, Kampus UI Depok, Depok 16424, Indonesia

E-mail: reyhan.dani@sci.ui.ac.id, hramad@ui.ac.id

Abstract. In this paper, we revisit Vilenkin's "*creation of universes from nothing*" in the framework of the Eddington-inspired Born-Infeld (EiBI) theory of gravity. In Vilenkin solution, a de Sitter space universe is created by quantum tunneling process from nothing. We incorporate the EiBI theory into this model. We use two kinds of cosmological constants, one from the EiBI theory and another from the effective potential. The resulting action has a slight modification than the original one.

1. Introduction

A modified theory of gravity was proposed by Banados and Ferreira in Ref. [1]. This new modified theory of gravity, called the Edington-inspired Born-Infeld (EiBI) theory, is built based on the original Eddington's proposal [2] by means of Born-Infeld-type of gravity [3]. This theory is constructed with Palatini formalism, where the metric and connection are independent with each other [4]. The formalism avoids various problems that plague the pure metric formalism, such as the ghost-like instabilities. With the absence of matter, the EiBI reduces to General Relativity. EiBI theory has been studied in wide variety of astrophysical and cosmological conditions, like in neutron stars [5], collapsing dust [6], black hole solutions [7, 8], and global monopole [9].

One major advantage of the EiBI theory is its ability to resolve the initial singularity of the universe [1]. If we consider the homogeneous and isotropic universe, we will find that there is a minimum length at early times. In classical sense, this describes a universe without an initial singularity.

Even though this result leads to a non-singular description of the universe, this does not necessarily mean that it does not have an origin point. In General Relativity, there are many proposals where the universe could be created by semi-classical method, such by quantum tunneling. One notable proposal is by Vilenkin [10, 11, 12]. In [10, 11] Vilenkin proposes that the universe is created from nothing to the de Sitter (dS) space by quantum tunneling using the Euclidean method. In the following paper [12], Vilenkin updates his result by proposing a different solution which uses Wheeler-DeWitt equations.

In this article, we would like to Vilenkin's first proposal into the EiBI theory. We would like to show that such a Vilenkin's proposal of creation from nothing using the Euclidean method is feasible within the EiBI framework. This result will further improve our understanding of the tunneling process using instanton in EiBI theory, thus this will help us when we want to explore

Content from this work may be used under the terms of the Creative Commons Attribution 3.0 licence. Any further distribution of this work must maintain attribution to the author(s) and the title of the work, journal citation and DOI.
Published under licence by IOP Publishing Ltd

another tunneling proposal. Here we work with the reduced Planck units $8\pi G = 1$ and set the speed of light $c = 1$, with an exception for section 3.

2. The Eddington-inspired born-infeld theory

The action in EiBI theory of gravity is given by [1],

$$S = \frac{1}{\kappa}\int d^4x \left(\sqrt{|g_{\mu\nu} + \kappa R_{\mu\nu}|} - \lambda\sqrt{|g|}\right) + S_M\left[g_{\mu\nu}, \Phi_M\right]. \tag{1}$$

$\lambda \equiv 1 + \kappa\Lambda$, where κ is the so-called Eddington parameter and Λ is the cosmological constant. S_M is the matter Lagrangian as a function of the metric $g_{\mu\nu}$ and the field Φ_M, not necessarily a scalar. The theory is constructed using the Palatini formalism, where the metric $g_{\mu\nu}$ and the connection Γ are treated as independent variables. As the result, the Ricci tensor $R_{\mu\nu}$ is purely a function of Γ. Only the symmetric part of the Ricci tensor $R_{\mu\nu}$ is considered.

Varying the action with respect to $g_{\mu\nu}$ and Γ, we obtain the field equations,

$$q_{\mu\nu} = g_{\mu\nu} + \kappa R_{\mu\nu}, \tag{2}$$

$$\sqrt{-q}q^{\mu\nu} = \lambda\sqrt{-g}g^{\mu\nu} - 8\pi\kappa\sqrt{-g}T^{\mu\nu}. \tag{3}$$

In the above equation $T_{\mu\nu}$ is defined as the energy-momentum tensor and $q_{\mu\nu}$ is an auxiliary metric. It should be noted that $q^{\mu\nu}$ is the inverse of $q_{\mu\nu}$, while raising and lowering index of $T_{\mu\nu}$ should be done with the physical metric $g_{\mu\nu}$.

The connection Γ is defined as,

$$\Gamma^{\mu}_{\alpha\beta} = \frac{1}{2}q^{\mu\nu}\left(q_{\nu\alpha,\beta} + q_{\nu\beta,\alpha} - q_{\alpha\beta,\nu}\right). \tag{4}$$

From above equation it should be clear that the connection Γ is not explicitly a function of the physical metric $g_{\mu\nu}$, but purely a function of the auxiliary metric $q_{\mu\nu}$

3. Vilenkin's creation of the universe from nothing proposal

In [10, 11], Vilenkin gives a cosmological model in which a de Sitter space universe is created by quantum tunneling process from nothing. In this model, we assume the universe starts in the symmetric vacuum state that has nonzero energy density ρ_v.The spacetime metric is given by

$$\mathrm{d}s^2 = \mathrm{d}t^2 - a^2(t)\mathrm{d}\Omega_3^2, \tag{5}$$

where $a(t)$ is the scale factor and $\mathrm{d}\Omega_3^2$ is the metric on a three-sphere. solution of $a(t)$ can be found from the evolution equation,

$$\dot{a}^2 + 1 = \frac{8}{3}\pi G\rho_v a^2, \tag{6}$$

where $\dot{a} = \mathrm{d}a/\mathrm{d}t$. The solution of above equation gives us the de Sitter space,

$$a(t) = H^{-1}\cosh(Ht), \tag{7}$$

where $H = (8\pi G\rho_v/3)^{1/2}$.

This solution describes a universe which is contracting at $t < 0$ until it reaches minimum size ($a_{min} = H^{-1}$) at $t = 0$, and is expanding at $t > 0$. This scenario is very similar to the particle's quantum tunneling scenario. Here we see a as the particle coordinate, in which this particle

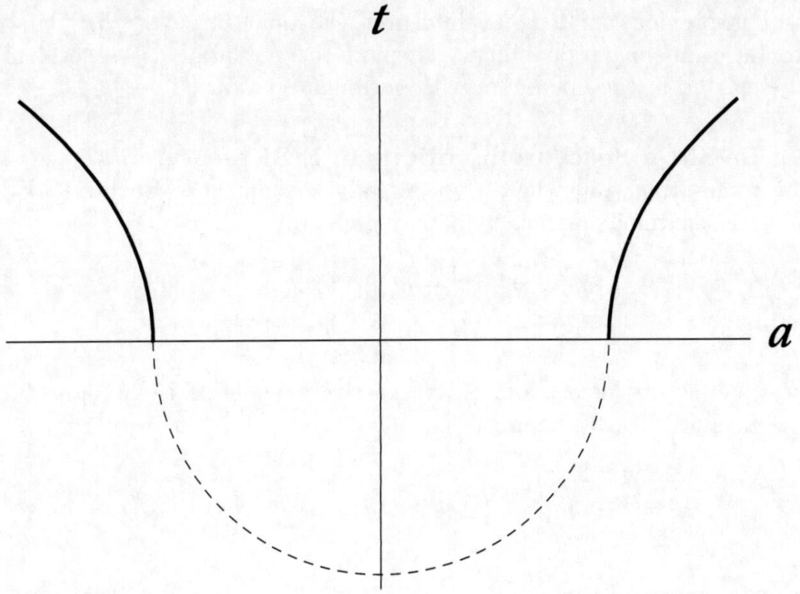

Figure 1. A schematic representation of the cosmic tunneling event.

bounces at a_{min}. This gives us a hint that the creation of the universe (in this case, only the de Sitter space universe) might be created from quantum tunneling event.

To discuss this cosmic tunneling event, we move to the Euclidean framework by Wick-rotating the time coordinate, $t \to -i\tau$. As a result, Eq. (6) becomes $-\dot{a}^2 + 1 = H^2 a^2$. Now the dot indicates the derivative with respect to τ. The solution of this equation is

$$a(\tau) = H^{-1}\cos(H\tau), \tag{8}$$

This solution also bounces at the classical turning point a_{min}. However, this solution did not approach any initial state at $\tau \to \pm\infty$, as now the solution is defined only for $|\tau| \leq \pi/2H$.

The tunneling process is illustrated in Fig. 1. The area below x-axis ($t < 0$) describes the universe before the nucleation. In this area, the universe is still 'living' in the Euclidean space. At $t = 0$, The universe undergoes a tunneling process from the Euclidean space to the Lorentzian dS space with $a = a_{min}$. Afterward, the universe continued to expand, illustrated by the area above x-axis ($t > 0$). We can interpret the tunneling process as a tunneling to dS from nothing, where 'nothing' means a state without classical Lorentzian spacetime.

We can further describe this tunneling process by calculate the action. The action of the de Sitter instanton is given by,

$$S_E = -\frac{1}{16\pi G}\int \mathrm{d}^4x R\sqrt{g} + \int \mathrm{d}^4x\sqrt{g}\left[\frac{1}{2}(\partial_\mu\phi)^2 + V(\phi)\right], \tag{9}$$

Where $V(\phi) = \rho_v$. We assume a slowly changing field $\phi \approx$ constant, thus removing the dynamics of ϕ. Using the solution of a, we can find the curvature of the instanton solution $R = 12H^2$. Under normal condition, we integrate the time from $-\infty$ to ∞. But we will only integrate from $-\pi/2H$ to $\pi/2H$, as the solution is only defined for that range. We find,

$$S_E = -\frac{3}{8G^2\rho_v}. \tag{10}$$

We can use our previous analogy to interpret the quantity $\exp(-S_E)$ to be something proportional to the tunneling probability. Then, if the potential has several maxima, most universes will be created at the maximum with the smallest value.

4. Creation of De Sitter Space from nothing in EiBI theory

We will use the result from [13], where they already calculate the Euclidean EiBI action for general case. We use the mini superspace metrics given by,

$$\begin{aligned} \mathrm{d}s_g^2 &= \mathrm{d}\tau^2 + a(\tau)^2 \mathrm{d}\Omega_3^2, &(11)\\ \mathrm{d}s_q^2 &= M(\tau)^2 \mathrm{d}\tau^2 + b(\tau)^2 \mathrm{d}\Omega_3^2, &(12)\end{aligned}$$

a and b are the scale factors for $g_{\mu\nu}$ and $q_{\mu\nu}$ respectively, while M is the lapse function. With this metrics, the Euclidean EiBI action is,

$$S_E = \frac{2\pi^2}{\kappa} \int \mathrm{d}\tau a^3 \left(\lambda + \kappa V(\phi) + \frac{\kappa\dot{\phi}^2}{2^2}\right)\left(1 - \frac{b^2}{a^2}\right), \tag{13}$$

with the equations of motion,

$$\begin{aligned} \ddot{b} &= \frac{b}{3\kappa} + \frac{\dot{b}\dot{M}}{M} - \frac{bM^2}{3\kappa}, &(14)\\ \frac{\dot{b}^2}{M^2} &= 1 + \frac{a^2}{2\kappa} - \frac{b^2}{3\kappa} - \frac{b^2}{6\kappa M^2}, &(15)\\ \frac{b^3}{a^3 M} &= \lambda + \kappa V(\phi) - \frac{\kappa\dot{\phi}^2}{2}, &(16)\\ \frac{bM}{a} &= \lambda + \kappa V(\phi) + \frac{\kappa\dot{\phi}^2}{2}, &(17)\\ 0 &= \ddot{\phi} + \dot{\phi}\frac{3\dot{a}}{a} - \frac{\mathrm{d}V(\phi)}{\mathrm{d}\phi}. &(18)\end{aligned}$$

These are the general results of [13] where we set $N(\tau) = 1$ (here $N(\tau)$ is the lapse function for $g_{\mu\nu}$).

Before we proceed, we can see that there are two kinds of the cosmological constants in this action, one coming from the minima of the potential, $V(\phi)$, and another one from the EiBI theory, embedded in $\lambda = 1 + \kappa\Lambda$. We will use these two simultaneously. Now we follow Vilenkin's approach as described in section 3. First, we set the scalar field ϕ to a constant value, thus removing the dynamics of the scalar field, $\dot{\phi} = 0$.

From the equations of motion we obtain

$$M^2 = \frac{b^2}{a^2} = \pm(\lambda + \kappa V(\phi)). \tag{19}$$

Note that the last part of above equation is a constant, as we assume that $\dot{\phi} = 0$ and λ is not dependent to τ. The other equations of motion will become,

$$\begin{aligned} \dot{a}^2 &= 1 + \frac{a^2}{3\kappa}(1 - (\lambda + \kappa V(\phi))), &(20)\\ \ddot{a} &= \frac{a^2}{3\kappa}(1 - (\lambda + \kappa V(\phi))). &(21)\end{aligned}$$

The solution for these equations is,

$$a = \sqrt{\frac{3}{(V(\phi)+\Lambda)}} \cos\left(\sqrt{\frac{(V(\phi)+\Lambda)}{3}}\tau\right). \tag{22}$$

As in the Vilenkin proposal, this solution is defined only for $|\tau| \leq (\pi/2)\sqrt{3/(V(\phi)+\Lambda)}$.

Now we can calculate the Euclidean EiBI action. We find,

$$\begin{aligned} S_E &= \frac{2\pi^2}{\kappa}\int d\tau a^3(\lambda+\kappa V(\phi))\left(1-\frac{b^2}{a^2}\right), \\ &= -2\pi^2(\lambda+\kappa V(\phi))(V(\phi)+\Lambda)\left(\frac{3}{V(\phi)+\Lambda}\right)^{3/2}, \\ &\times \int_{-\frac{\pi}{2}\sqrt{\frac{3}{(V(\phi)+\Lambda)}}}^{\frac{\pi}{2}\sqrt{\frac{3}{(V(\phi)+\Lambda)}}} d\tau \cos^3\left(\sqrt{\frac{(V(\phi)+\Lambda)}{3}}\tau\right), \\ &= -8\pi^2(\lambda+\kappa V(\phi))\frac{3}{(V(\phi)+\Lambda)}, \\ &= -24\pi^2\left(\frac{1}{(V(\phi)+\Lambda)}+\kappa\right). \end{aligned} \tag{23}$$

Note that we work with the reduced Planck units $8\pi G = 1$. This result only differs from the result in the previous section by the extra κ term and the combined cosmological constant.

The value of this action could be changed by using a different value of κ. By using a positive value of κ, the action will be more negative than the general relativity counterpart. The opposite situation happens if we use a negative value of κ, the action will be more positive. Also, the action could cross $S_E = 0$ and change sign to a positive value for the bigger value of $(V(\phi)+\Lambda)$ when we use a negative value of κ.

Another point here, as there are two types of cosmological constants we can choose to use only one of them by setting the other to zero. The resulting action would be same, but it can have a different initial condition. If we choose to set $V(\phi) = 0$, the action will be independent of ϕ. Surprisingly, even if we remove the contribution from the scalar field and the potential, we still arrive at the same action. Hence, the resulting action comes purely from gravity.

In [1], Bañados and Ferreira found that for $\kappa > 0$, the universe loiters for a very long time before it expands to the usual cosmological evolution. If we rewind back the time, the scale factor will need an infinitely long time before reaching the minimum length a_B. This behavior describes an universe without an initial singularity, because there will be no initial point for the universe. But it should be noted that all calculation in [1] is done classically. The universe might be created by a semi-classical process. Let us assume that at the moment of creation, the universe tunneled directly to, or near, the minimum length, $a_B \sim 10^{-32}(\kappa)^{1/4}a_0$, where a_0 is the scale factor today. Taking the amplitude of (22) as a_B, we can find the Euclidean action for tunneling to a_B as $-24\pi^2\left(10^{-64}\kappa^{1/2}a_0^2/3+\kappa\right)$. This gives a possible explanation for a universe with non-singular beginning in EiBI theory.

5. Conclusion

In this article, we incorporate Vilenkin's creation of the universes from nothing into EiBI theory. We show that such proposal is feasible within the EiBI framework; i.e., the universe governed by the EiBI gravity could have been created from nothing by means of quantum tunneling. We calculate the corresponding Euclidean action. If we use a positive value of Eddington parameter

κ, the Euclidean action is decreased. Otherwise, it is increased. We also find that with EiBI theory, the Euclidean action can be achieved without the scalar field contribution, resulting in action that comes only from gravity.

In our future works, we would like to explore another Instanton solution and incorporate it into EiBI theory. Also we want to try the feasibility of Vilenkin's another proposal [12] that uses Wheeler-DeWitt equations in EiBI theory.

Acknowledgment
We thank Jason Kristianto for the useful discussion. This work is partially funded by PITTA grant from Universitas Indonesia under contract no. 2267/UN2.R3.1/HKP.05.00/2018.

References
[1] M. Banados and P. G. Ferreira, "Eddington's theory of gravity and its progeny," Phys. Rev. Lett. **105** (2010) 011101
[2] A. S. Eddington, "The Mathematical Theory of Relativity," Cambridge University Press (1930)
[3] M. Born and L. Infeld, "Foundations of the new field theory," Proc. Roy. Soc. Lond. A **144**, no. 852, 425 (1934).
[4] D. N. Vollick, "Palatini approach to Born-Infeld-Einstein theory and a geometric description of electrodynamics," Phys. Rev. D **69** (2004) 064030
[5] P. Pani, V. Cardoso and T. Delsate, "Compact stars in Eddington inspired gravity," Phys. Rev. Lett. **107** (2011) 031101
[6] T. Delsate and J. Steinhoff, "New insights on the matter-gravity coupling paradigm," Phys. Rev. Lett. **109** (2012) 021101
[7] S. Jana and S. Kar, "Born-Infeld gravity coupled to Born-Infeld electrodynamics," Phys. Rev. D **92** (2015) 084004
[8] B. N. Jayawiguna and H. S. Ramadhan, "Charged black holes in higher-dimensional Eddington-inspired Born-Infeld gravity," arXiv:1810.08780 [gr-qc].
[9] R. D. Lambaga and H. S. Ramadhan, "Gravitational field of global monopole within the Eddington-inspired Born-Infeld theory of gravity," Eur. Phys. J. C **78** (2018) no.6, 436
[10] A. Vilenkin, "Creation of Universes from Nothing," Phys. Lett. **117B**, 25 (1982).
[11] A. Vilenkin, "The Birth of Inflationary Universes," Phys. Rev. D **27** (1983) 2848.
[12] A. Vilenkin, "Quantum Creation of Universes," Phys. Rev. D **30**, 509 (1984). doi:10.1103/PhysRevD.30.509
[13] F. Arroja, C. Y. Chen, P. Chen and D. h. Yeom, "Singular Instantons in Eddington-inspired-Born-Infeld Gravity," JCAP **1703** (2017) no.03, 044

Deformation of quark stars under strong internal magnetic fields

R. Rizaldy, A. Sulaksono*

Departement of physics, FMIPA, University of Indonesia, Kampus UI Depok, Indonesia

E-mail: `ryan.rizaldy@sci.ui.ac.id` and `*anto.sulaksono@sci.ui.ac.id`

Abstract. The pulsar detector has observed the existence of a strong magnetic field on compact stars. So much we can learn from this observation, one of them is the deformation of a compact star under strong magnetic fields. In this note, we study the deformation of quark stars due to a magnetic field. Note that quark stars and neutron stars have a fundamental difference in their equation of states such as the composition and crusts. Here, we used two equation of states of quark matter, i.e., CIDDM model with the scalar and with vector Coulomb terms included [1]. In both equation of states, we vary the magnetic field value in the center of the star to see how the internal magnetic field affects the shape of the quark star.

1. Introduction

After Anomalous Detectors X-ray pulsars (AXPs) and Soft-Gamma Repeaters (SGR) detect strong magnetic fields present in a pulsar where reported as remnants of supernovae are classified as young neutron stars [2, 3], many researchers are interested in studying effect of magnetic fields on relativistic stars. The observations from AXPs detect the magnetic field value at the star surface of about $10^{14} - 10^{15}$ G. The results of SGR and AXPs observations are believed to be magnetars i.e. neutron stars with very strong magnetic fields [4–7]. But as we know that neutron stars are not the only product of supernova explosions, but there are types of stars that have enormous mass densities that can break down neutrons into their quark-quark to form quark stars [8]. Quark stars are hypothesis stars that contain three quarks, namely up, down, and strange. The up and down quarks themselves are formed because of the very strong gravitational pressure and are able to break down neutrons in neutron stars into their constituents, up and down quarks. Strange quarks are formed from converted of up and down quarks through weak interactions [8, 9].

There are two ways to give the magnetic field effect to the relativistic star, first on the microscopic aspect, that is, the presence of a strong magnetic field is found in the energy levels of charged particles caused by Landau quantization, thus modifying the equation of state (EoS) for matters in stars, researchers have used this method a lot and apply it to EoS relativistic stars i.e. hyperon stars [10, 12], quark stars [11] and hybrid stars [13]. The second is by macroscopic means where the magnetic field effect is added into tensor energy-momentum as perturbative term, as which is done by Ref. [16, 17] (see also the references therein).

We can apply the deformations that occur in relativistic stars using Hartle and Thorne approximation [14, 15], which is a multipole expansion of the schwarzschild metrics, and pair it

Content from this work may be used under the terms of the Creative Commons Attribution 3.0 licence. Any further distribution of this work must maintain attribution to the author(s) and the title of the work, journal citation and DOI.
Published under licence by IOP Publishing Ltd

with the sum of the magnetic field at the energy-momentum tensor in such a way as to cause the deformation [16, 17].

2. Formalism Detail

We have solved the equations that Hartle and Thorne have derived from their paper, which are used to describe the deformation of slow-motion rotation on relativistic stars [14, 15]. In this study, we assume that the magnetic field that causes deformation in which our formulation is exactly the same as that done on Ref. [16, 17], where Hartle and Thorne's perturbation metrics are expressed as follows

$$\begin{aligned} ds^2 = &- e^{\nu}\left[1 + 2\left(h_0 + h_2 P_2(\cos\theta)\right)\right] dt^2 \\ &+ e^{\lambda}\left[1 + \frac{e^{\lambda}}{r}\left(m_0 + m_2 P_2(\cos\theta)\right)\right] dr^2 \\ &+ r^2\left[1 + 2k_2 P_2(\cos\theta)\right]\left(d\theta^2 + \sin^2\theta d\phi^2\right), \end{aligned} \quad (1)$$

where m_0, m_2, h_0, h_2 and k_2 is a function of r and $P_2(\cos\theta)$ is Legendre function. The perturbation metrics have three summations, the first part is the Schwarzschild metric, the second part with the subscript '0' is the monopole correction and the last with subscript '2' is the quadrupole correction. The perturbation metrics need to be solved separately by connecting them to the energy-momentum tensor which also undergoes expansion in the magnetic field, which is formulated as follows

$$\varepsilon = \varepsilon_m + \frac{B^2}{8\pi} \quad (2)$$

$$P_{\perp} = P_m + \frac{B^2}{8\pi} \quad (3)$$

$$P_{\parallel} = P_m - \frac{B^2}{8\pi}, \quad (4)$$

where ε_m and P_m is the energy density and thermodynamics pressure of matter, $B^2/8\pi$ is the magnetic stress along z-axis, Total pressure can be written in Legendre polynomial form

$$P = P_m + p_0 + p_2 P_2(\cos\theta) \quad (5)$$

$$\varepsilon = \varepsilon_m + 3p_0, \quad (6)$$

where $p_0 = B^2/24\pi$ is the monopole factor and $p_2 = -4B^2/24\pi$ is the quadrupole factor.

We find the following equations after solving Einstein's field equations:

$$\frac{dP}{dr} = -\frac{G}{r^2}\frac{(\varepsilon_m + P_m)(m + 4\pi r^3 P_m)}{(1-2Gm/r)} \tag{7}$$

$$\frac{dm}{dr} = 4\pi r^2 \varepsilon_m \tag{8}$$

$$\frac{dm_0}{dr} = 12\pi r^2 p_0 \tag{9}$$

$$\frac{dh_0}{dr} = 4\pi G r e^{\lambda} p_0 + \frac{1}{r} G \nu' e^{\lambda} m_0 + \frac{1}{r^2} G e^{\lambda} m_0 \tag{10}$$

$$\frac{d}{dr}(h_2 + k_2) = h_2\left(\frac{1}{r} - \frac{1}{2}\frac{d\nu}{dr}\right) + \frac{e^{\lambda}}{r} G m_2 \left(\frac{1}{r} + \frac{1}{2}\frac{d\nu}{dr}\right) \tag{11}$$

$$h_2 + \frac{e^{\lambda}}{r} G m_2 = 0 \tag{12}$$

$$\frac{dh_2}{dr} + \frac{dk_2}{dr}\left(1 + \frac{r}{2}\frac{d\nu}{dr}\right) = 4\pi G r e^{\lambda} p_2 + \frac{e^{\lambda}}{r^2} G m_2 + \frac{e^{\lambda}}{r} G \frac{d\nu}{dr} m_2 + \frac{3}{r} e^{\lambda} h_2 + \frac{2}{r} e^{\lambda} k_2, \tag{13}$$

we also get the following equations from Bianchi's contracted identity:

$$\frac{d\nu}{dr} = -\frac{2}{(\varepsilon_m + P_m)}\frac{dP}{dr} \tag{14}$$

$$\frac{dp_0}{dr} = -2\frac{d\nu}{dr} p_0 - (\varepsilon_m + P_m)\frac{dh_0}{dr} \tag{15}$$

$$p_2 = -(\varepsilon_m + P_m)\, h_2 \tag{16}$$

$$\frac{dp_2}{dr} = -\frac{1}{2}\frac{d\nu}{dr} p_2 - (\varepsilon_m + P_m)\frac{dh_2}{dr}. \tag{17}$$

Equation (7) - (17) selected and simplified in such a way as to be solved using numerical methods, provided that EoS of a relativistic star is known, in this case it is a quark star. The energy density equations of the materials we use are based on the author's research of Ref. [1], where in that paper the material density energy is applied with CIDDM model with Lagrangian as follows

$$\mathcal{L}^{Int(CIDDM)} = -\kappa_1 n_B^{1/3} n_B^{(s)} - \kappa_3 \delta n_B^{a} e^{-b n_B} n_B^{(s\tau)}, \tag{18}$$

where κ_1 s the independent isospin quark parameter, κ_3 is a quark dependent interaction parameter, a and b are isospin parameters in the CIDDM model, $\delta = 3\,(n_d - n_u)\,/\,(n_d + n_u)$ is an up-down quark asymmetry parameter. Then $n_b, n_b^{(s)}$ and $n_B^{(s\tau)}$ are vector-isoscalar, scalar-isoscalar and vector-isovector baryon number density respectively. From Ref. [1], they focus on investigating Lagrangian density interactions

$$\mathcal{L} = \mathcal{L}^{Int(CIDDM)} + \Delta\mathcal{L}^{(i)Int}, \tag{19}$$

where for $i = 1$ using Coulomb scalar density model,

$$\Delta\mathcal{L}^{(1)Int} = -\kappa_2^{(1)} n_B^{1/3} n_B^{(s)}, \tag{20}$$

and for $i = 2$ using Coulomb vector density model,

$$\Delta\mathcal{L}^{(2)Int} = -\kappa_2^{(2)} n_B^{4/3}, \tag{21}$$

research results from Ref. [1] found a stable condition on $\kappa_2^{(1)} = 0.3$ for range value $0.45 \lesssim \kappa_1 \lesssim 0.54$, and $\kappa_2^{(2)} = -0.8$ for range value $0.63 \lesssim \kappa_1 \lesssim 0.83$. In this paper we use EoS with degrees of freedom κ_1 and κ_2.

The magnetic field term B obtained using ansatz from Ref. [10] sebagai berikut

$$B\left(n_B/n_0\right) = B_{surf} + B_0\left[1 - exp\{-\beta\left(n_B/n_0\right)^{\gamma}\}\right], \tag{22}$$

with assumption that the condition of magnetic field in the center of the star that must be higher than the surface. we set the value of saturation density $n_0 = 0.15 fm^{-3}$. Here we set at $\beta = 0.01$ and $\gamma = 2$ according to Ref. [10]. the magnetic field at the surface is $B_{surf} = 10^{15}$ G, which is according to the observations data and $B_0 = 5 \times 10^{18}$ also obtain from Ref. [10].

3. Result and Discussion

The result of the numerical calculation we get the values of $P, m, r, m_0, m_2, h_0, h_2$ and k_2, in this study we focus more on the deformation that occurs due to the presence of magnetic field. First we can define the value of mass and the new radius due to the presence of correction factor monopole and quadrupole, that is for total mass

$$M = m + m_0, \tag{23}$$

where m is the spherical mass and m_0 is mass for corrections factor to the magnetic field. We can write the radius equatorial and polar according to Ref. [14–16]

$$R_e = R + \xi_0(R) - \frac{1}{2}\left(\xi_2(R) + Rk_2(R)\right) \tag{24}$$

$$R_p = R + \xi_0(R) + \left(\xi_2(R) + Rk_2(R)\right), \tag{25}$$

where R radius of spherical stars, ξ_0 and ξ_2 define by

$$\xi_0(r) = -h_0\frac{r(r - 2Gm(r))}{G(4\pi r^3 P_m + m(r))} \tag{26}$$

$$\xi_2(r) = -h_2\frac{r(r - 2Gm(r))}{G(4\pi r^3 P_m + m(r))}. \tag{27}$$

We can define degrees of deformation by using the equation of ellipticity

$$e = \sqrt{1 - \chi^2}, \tag{28}$$

where to prolate shape we define $\chi = R_e/R_p$ and for oblate shape $\chi = R_p/R_e$.

Figure 1 illustrates the relation between mass and star radius using two different models. In panel (a) of Fig. 1, we use the Coulomb scalar EOS CIDDM model with two values $\kappa_1 = 0.54$ and 0.45 which are the representation of the equation of states with minimum and maximum stiffness allowed by the stability of two flavors strange matter equation of state constraint. It can be seen in both models that after applying the magnetic field that except the region close to maximum mass, the changes due to magnetic fields are not too significant. It is evident that around the maximum mass the relation of $R_p > R_e$ is fulfilled, while outside this region, on the contrary, $R_e > R_p$ is fulfilled by both of κ_1 values. Similar situation but for Coulomb vector can be also seen in panel (b) of Fig.1. In that case, we can observe that the quark star is slightly deformed into the prolate form around the maximum mass region this happens due to the assumption that the magnetic field direction puts in the z-axis.

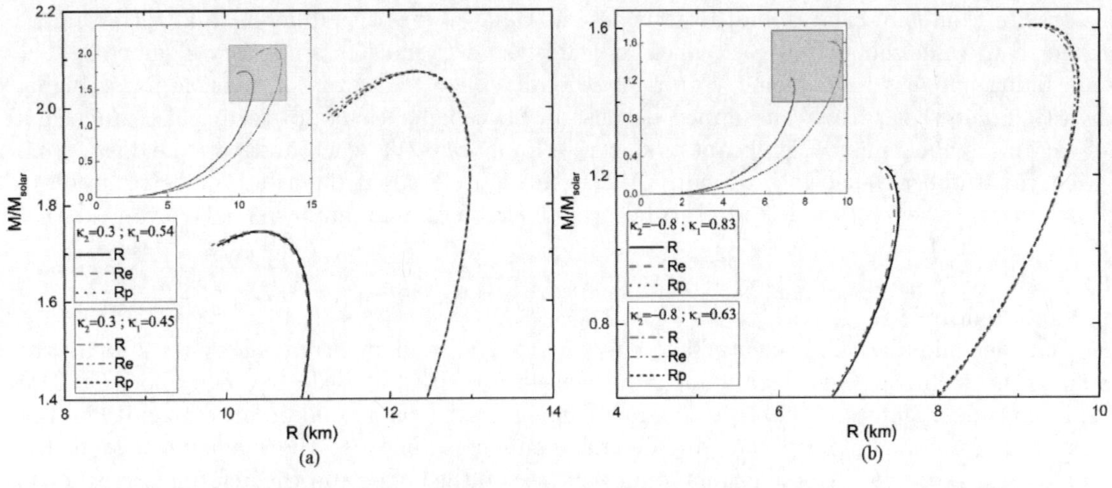

Figure 1. Relation between total mass and radius of quark stars according CIDDM model

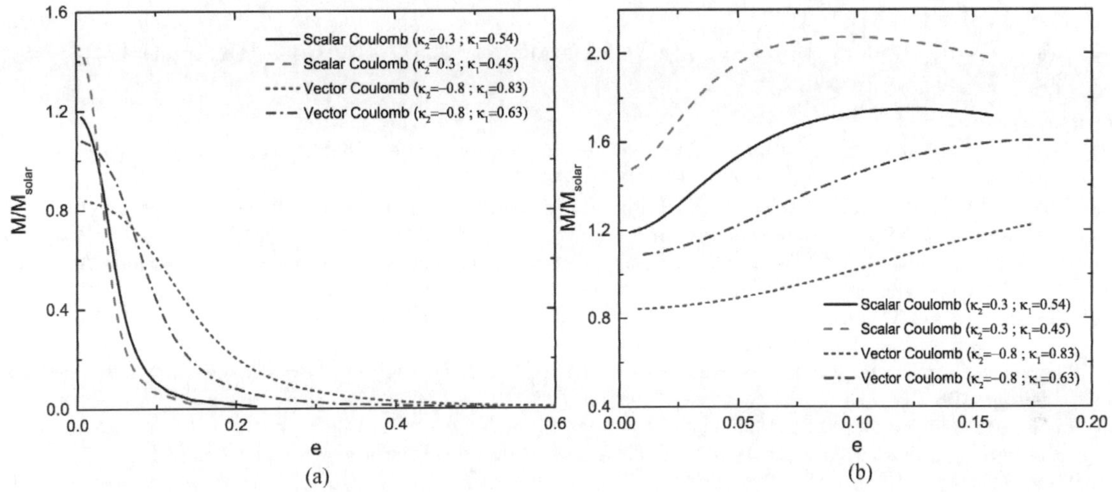

Figure 2. Relation between ellipticity and Mass, (a) for $\chi = R_p/R_e$ and (b) for $\chi = R_e/R_p$.

Table 1. Profile of quark stars on the maximum mass in each CIDDM model.

Model	M/M_{solar}	R/km	Re/km	Rp/km	e
Coulomb scalar ($\kappa_1 = 0.54$)	1.746	10.427	10.375	10.455	0.123
Coulomb scalar ($\kappa_1 = 0.45$)	2.073	12.336	12.298	12.355	0.096
Coulomb vector ($\kappa_1 = 0.83$)	1.223	7.324	7.247	7.361	0.175
Coulomb vector ($\kappa_1 = 0.63$)	1.606	9.226	9.135	9.282	0.176

Figure 2 is the relation between the mass and the ratio of the ellipticity of quark stars where with the range of values $e \sim 0 - 1$, 0 indicating no deformation at all. From Figure 2 (a) is the degree of deformation of the oblate form, in which stars tend to be oblate-shaped in small masses. In Coulomb scalar, in both variations the value κ_1 is not so different that the e value is in the small value range ie $0 - 0.2$ which indicates the deformation is not very significant. The same thing happens to Coulomb vector but the difference of degrees of deformation is higher than Coulomb scalar, this can happen because in Lagrangian Coulomb vector the influence of its baryon number is more significant as shown in Equation (21), where magnetic field is depend on baryon number as shown in Equation (22). In Figure 2 (b) it describes for deformation to prolate form, where all models tend to be prolate shape on maximum mass. We can see that detail value in Table 1.

4. Conclusion

We can conclude that the magnetic field effect added to the momentum energy tensor with the value of the magnetic field constant at the surface of the quark star $B_0 = 10^{15}$ G for both CIDDM models has less significant effects (the radius difference exists in second order of magnitude) but make the star tend to be prolate in the maximum mass because the magnetic field in the neutron star is on the z-axis to form a repulsive force on the z-axis and the deformation is not so significant possible because of the gravitational force dominates on a high mass of quark stars.

ACKNOWLEDGMENT

We are partially supported by the UIs PITTA grant No. 2231/UN2.R3.1/HKP.05.00/2018.

References

[1] Qauli A I and Sulaksono A 2016 Quark matter at high density based on extended confined isospin-density-dependent mass model *Physical Review D* **93** 104056
[2] Kulkarni S R and Frail D A 1993 Identification of a supernova remnant coincident with the soft gamma-ray repeater SGR1806-20 *Nature (London)* **365** 33
[3] Murakami T, Tanaka Y, Kulkarni S R, Ogasaka Y, Sonobe T, Ogawara Y, Aoki T,p and Yoshida A, 1994 X-ray identification of the soft -ray repeater 1806 20 *Nature* **368** 127
[4] Duncan R C and Thompson C 1992 Formation of very strongly magnetized neutron stars - Implications for gamma-ray bursts, *AstroPhysical Journal*, **392** L9
[5] Thompson C and Duncan R C 1993 Neutron star dynamos and the origins of pulsar magnetism *Astrophysical Journal* **408** 194
[6] Thompson C and Duncan R C 1995 The soft gamma repeaters as very strongly magnetized neutron stars - I. Radiative mechanism for outbursts *Monthly Notices of the Royal Astronomical Society* **275** 255
[7] Thompson C and Duncan R C 1996 The Soft Gamma Repeaters as Very Strongly Magnetized Neutron Stars. II. Quiescent Neutrino, X-Ray, and Alfven Wave Emission *Astrophysical Journal* **473** 322
[8] Weber F 2005 Strange Quark Matter and Compact Stars *Progress in Particle and Nuclear Physics* **54** 193
[9] Witten E 1987 Cosmic separation of phases *Physical Review D* **30** 272
[10] Chakrabarty S, Bandyopadhyay D and Pal S 1997 Dense nuclear matter in a strong magnetic field *Physical Review Letter* **78** 2898.
[11] Sinha M, Mukhopadhyay B and Sedrakian A 2013 hypernuclear matter in strong magnetic field *Nuclear Physics* **A898** 43
[12] Isayev A A 2015 Anisotropic pressure in the quark core of a strongly magnetized hybrid star, Journal of Physics: Conference Series *607* 012013
[13] Dexheimer V, Negreiros R, Schramm S and Hempel M 2013 Deconfinement to quark matter in neutron stars the influence of strong magnetic fields *AIP Conference Proceeding* **1520** 264
[14] Hartle J B 1967 Slowly Rotating Relativistic Stars *Astrophyical Journal* **150** 1005
[15] Hartle J B and Thorne K S 1968 Slowly Rotating Relativistic Stars. II. Models for Neutron Stars and Supermassive Stars *Astrophysical Journal* **153** 807
[16] Mallick R and Schramm S 2014 Deformating of a magnetized neutron star *Physical Review C* **89** 045805-1
[17] Rizaldy R and Sulaksono A 2018 Solution of Einsteins Equation for Deformation of a Magnetized Neutron Star *Journal of Physics: Conference Series* **1011** 012083

Matter perturbation in coupled scalar field cosmology

F A M Mulki and H Wulandari

Astronomy Research Division, Faculty of Mathematics and Natural Sciences, Institut Teknologi Bandung, Jalan Ganesha 10, Bandung 40132, Indonesia

fargizaabdangd@gmail.com

Abstract. A universe containing a coupled scalar field would have different dynamical evolution compared to uncoupled cases. Significant differences are expected to emerge during matter dominated era, where scalar field density mimics how matter density evolves, by becoming subdominant. Analysis of the dynamics was investigated both analytically and numerically through phase plane method, and we obtained two attractor solutions which are compatible to a late time cosmic acceleration as the ending of matter dominated era. Ordinary scalar field triggers an acceleration with stable attractor solution which converges to $\omega_\phi = \omega_{eff} \approx -1$. Whereas, phantom scalar field allows negative values for its own energy ($\omega_{eff} < -1$ and $\Omega_\phi < 0$); however this is not stable, therefore it will be caught by attractor solution which is stable to $\omega_\phi = \omega_{eff} \approx -1$. In addition, we found cosmological parameters generated from both analytical and numerical calculations, i.e. ω_ϕ, Ω_ϕ and ω_{eff}, by assuming a flat universe. The existence of a coupling constant Q between scalar field and matter induces different structure growth compared to uncoupled cases and standard cold dark matter (CDM) model. During matter dominated era, ordinary scalar field induces growth of structures so that it becomes faster than the standard model does, whereas phantom scalar field slows down the growth of structures. During scalar field dominated era (far in the future), we obtained that formed structures or density contrasts will decay in a similar way as in uncoupled cases, whereas following phantom scenario, the decay of formed structures occurs faster and more significant. This result is associated to background dynamics which shows that big rip will be the relevant fate of our universe.

1. Introduction

One of latest issues in cosmology which is yet to solve is the cosmic acceleration [1,2]. Various proposals have been put forward to explain this problem, basically by searching new kind of energy or dark energy and by modifying the theory of gravity. The simplest and most popular scenario belongs to dark energy sector is the cosmological constant, which is regarded as vacuum energy stored in spacetime. However, two serious problems rise from this model which are dubbed as fine tuning and coincidence problem [3]. Alternative models have been conceived by assuming dark energy in the form of scalar fields. The simplest model of scalar fields is quintessence with various types of potential [4,5,6,7,8]. However, the observational facts that matter-energy density nowadays is the same in order of magnitude as dark energy density suggest the possibility of coupling between them, which is mediating energy to transform. This suggestion has motivated many authors to make a generalization from uncoupled into coupled scalar field.

The existence of coupling between matter and dark energy may exhibit different signatures of the universe, especially in higher order of dynamics, i.e. inhomogeneity and anisotropy of the universe. Study of scalar field coupled to matter and its effects to structures growth have been reported well by

Content from this work may be used under the terms of the Creative Commons Attribution 3.0 licence. Any further distribution of this work must maintain attribution to the author(s) and the title of the work, journal citation and DOI.
Published under licence by IOP Publishing Ltd

different authors [9,10,11,12,13]. In this paper we adopt the model [13] and propose another technique and approach to formulate the perturbation equation analytically and find solution for its dynamics through stability analysis of its critical points. Regarding the results from [14] that allows possibility for $\omega_\phi < -1$, we consider here a general scalar field which includes phantom field with negative kinetic energy.

2. Background dynamics

An anzats lagrangian for scalar field coupled to matter is

$$L(X,\phi) = X - V(\phi) - Q\rho_m\phi \tag{1}$$

where Q, ρ_m and ϕ denote coupling constant, general matter density and scalar field respectively, and $X = -\frac{1}{2}g^{\mu\nu}\nabla_\mu\phi\,\nabla_\mu\phi$ is the kinetic component of the scalar field. For lagrangian in eq. (1) we can derive the equation of motion for scalar field in Friedmann universe,

$$\phi'' + 2\mathrm{H}\phi' + a^2 V_{,\varphi} = -Q\rho_m. \tag{2}$$

which is usually known as Klein-Gordon equation. In the above equation, the accent sign stands for the derivative with respect to conformal time η which is related to cosmic time t by $dt = ad\eta$. a represents a scale factor whose value is normalized to 1 today ($a_0 = a(t_0) = 1$) and H is the Hubble parameter defined as $\mathrm{H} = a'/a$ in conformal time. The scalar field coupled to general matter satisfies the condition

$$\nabla_\mu T^\mu_\nu = \pm Q T_M \nabla_\nu\phi, \tag{3}$$

where the plus (+) sign refers to matter and radiation, minus (−) sign refers to scalar field, and T_M is the trace of fluid energy-momentum tensor. This condition yields a set of continuity equations as follows

$$\dot{\rho}_r + 4H\rho_r = 0,$$

$$\dot{\rho}_m + 3H\rho_m = Q\rho_m\dot{\phi},$$

$$\dot{\rho}_\phi + 3H\left(\rho_\phi + P_\phi\right) = -Q\rho_m\dot{\phi}, \tag{4}$$

where dot sign refers to the derivative with respect to cosmic time t and $H = \dot{a}/a$. The dynamics analysis of the coupled scalar field cosmology can be started by investigating the existence of scaling condition, which acts as the boundary between accelerating and decelerating modes.

2.1. General scaling condition

For simplification, we rewrite Friedmann equation in flat universe as $H^2 = \beta\rho_T$, where $\beta = \sqrt{8\pi G/3}$ and ρ_T is the total density of the universe. The general form of P_ϕ in general relativity framework is given by [15],

$$P_\phi = Xh\left(Xe^{\lambda\phi}\right), \tag{5}$$

where $h(Xe^{\lambda\phi})$ is a function of $Y \equiv Xe^{\lambda\phi}$. During scaling era it can be proven that $h(Xe^{\lambda\phi}) =$ constant $= Y_0$ [13]. The yield reveals that pressure is proportional to kinetic component and only determined by its kinetic term. To construct our dynamical system of the universe, one should be recognized is the evolution of scalar field density expressed in general variables. We can write a general form of scalar field density as

$$\rho_\phi = 2X\frac{\partial P_\phi}{\partial X} - P_\phi. \tag{6}$$

By using eq. (5), eq. (6) and their derivatives with respect to time we obtained

$$\rho_\phi = \left[1 + 2Xe^{\lambda\phi}\frac{\partial h\left(Xe^{\lambda\phi}\right)}{\partial\left(Xe^{\lambda\phi}\right)}\frac{1}{h\left(Xe^{\lambda\phi}\right)}\right]Xh\left(Xe^{\lambda\phi}\right).$$

$$\dot{\rho}_\phi = h(Y)\ddot{\phi}\dot{\phi} + 2h_{,Y}\ddot{\phi}\dot{\phi}^3 e^{\lambda\phi} + \frac{1}{2}h_{,Y}\dot{\phi}^5\lambda e^{\lambda\phi} + 2XYh_{,YY}\ddot{\phi}\dot{\phi}e^{\lambda\phi} + XYh_{,YY}\dot{\phi}^3\lambda e^{\lambda\phi}. \tag{7}$$

Then, by setting $8\pi G = 1$ and doing a little mathematics to eq. (4), and (7) we obtained

$$\left[h(Y) + 4Yh_{,Y} + 2Y^2 h_{,YY}\right]\ddot{\phi} + 3H\left[h(Y) + Yh_{,Y}\right]\dot{\phi} + 2XY\lambda h_{,Y} + 2XY^2\lambda h_{,YY} = -Q\rho_m,$$

which is the scaling Klein-Gordon equation. Next, by introducing dimensionless variables [12]

$$x \equiv \frac{\dot{\phi}}{\sqrt{6}H}, \quad y \equiv \frac{e^{-\lambda\phi/2}}{\sqrt{3}H},$$

and using several aforementioned equations, we can construct a system that describes our picture of the universe. However, for mathematical requirements, we propose an assumption that would simplify our analytical treatment, by assuming a scalar field which fulfill the condition

$$\frac{1}{2}\lambda\dot{\phi}^2 = -\ddot{\phi}.$$

Under this condition the dynamics of the scalar field is governed by a set of equations:

$$\frac{dx}{dN} = x\left[-\frac{1}{H}\frac{dH}{dN} - \frac{3\{h(Y) + Yh_{,Y}\}}{2Y^2 h_{,YY} + 5Yh_{,Y} + h(Y)}\right] - \sqrt{6}\frac{\lambda x^2 Y\{3h_{,Y} + 2Yh_{,YY}\} + Q\left[1 - x^2\{h(Y) - 2Yh_{,Y}\}\right]}{2\left(2Y^2 h_{,YY} + 5Yh_{,Y} + h(Y)\right)}.$$

$$\frac{dy}{dN} = -\frac{\sqrt{6}}{2}\lambda xy - y\frac{1}{H}\frac{dH}{dN}, \tag{8}$$

$$\frac{1}{H}\frac{dH}{dN} = -\frac{3}{2}\left[1 + \omega_m + (1-\omega_m)x^2 h(Y) - 2\omega_m x^2 Yh_{,Y}\right],$$

where ω_m and ω_ϕ are state parameters for matter and scalar field respectively, and $\omega_{eff} = \omega_\phi\Omega_\phi + \omega_m\Omega_m$. The above set of equations is an autonomous system, whose solutions are dependent on the explicit form of $h(Y)$.

An autonomous system is a class of dynamical system that plays important role in cosmology [18] and contains two or more coupled differential equations. In eq. (8), there are two coupled differential quations in the form $x = f(x, y, N)$ and $y = g(x, y, N)$ and each of them do not contain explicit $N-$ dependent terms. The critical points of the system can be found by fixing $dx/dN = 0$ and $dy/dN = 0$ simultaneously.

2.2. *The dynamics of scalar field with* $V = V_0 e^{-\lambda\phi}$

This section deals with specific scalar fields that naturally has scaling solutions. An exponential potential is a natural form of scalar field potentials, meaning that it is an exact and simpler solution of Klein-Gordon equations. Thus, we write the lagrangian as below

$$L(X,\phi) = P_\phi = \varepsilon X - V_0 e^{-\lambda\phi}, \tag{9}$$

where $\varepsilon = +1$ for ordinary scalar field and $\varepsilon = -1$ for phantom scalar field. Lagrangian (9) is assosiated with $h(Xe^{\lambda\phi}) = \varepsilon - V/Xe^{\lambda\phi}$, consequently, the autonomous equations(8) becomes

$$\frac{dx}{dN} = -3x + \frac{\sqrt{6}}{2\varepsilon}\lambda V_0 y^2 + \frac{3}{2}x\left[(1-\omega_m)\varepsilon x^2 + (1+\omega_m)(1 - V_0 y^2)\right] - \frac{\sqrt{6}}{2\varepsilon}Q\left(1 - \varepsilon x^2 - V_0 y^2\right),$$

$$\frac{dy}{dN} = -\frac{\sqrt{6}}{2}\lambda xy + \frac{3}{2}y\left[(1-\omega_m)\varepsilon x^2 + (1+\omega_m)(1 - V_0 y^2)\right].$$

These give critical points, whose relations to cosmological parameter are resumed in Table 1, and in the phase diagrams that show the arrows of stability in Figure 1 and 2.

Journal of Physics: Conference Series **1354** (2019) 012007 doi:10.1088/1742-6596/1354/1/012007

2.3. *Stability and dynamics analysis*

2.3.1. Ordinary scalar field ($\varepsilon = +1$). Based on its energy, point (a) is a kinetic dominated era, $\omega_\phi = 1 > -1/3$ and $\omega_{eff} > -1/3$ since the second term, i.e. $\frac{2}{3}\frac{Q^2}{1-\omega_m} > 0$ and $\omega_m > 0$, which corresponds to a deceleration condition. Points (b) & (c) belong also to kinetic and scalar field dominated eras, however these will never trigger an acceleration since $\omega_{eff} = \omega_\phi = 1 > -1/3$. Points (d) and (e) are scalar field dominated solutions ($\Omega_\phi = 1$) with $\omega_{eff} = \omega_\phi$, therefore acceleration occurs if $\lambda < \sqrt{2}$ is fulfilled. Points (f) & (g) are scaling solutions. Since Q, λ and ω_m are constants, hence $\Omega_\phi =$ constant and $\Omega_m = 1 - \Omega_\phi =$ constant, therefore $\frac{\Omega_\phi}{\Omega_m} =$ constant. The requirement for $\Omega_\phi < 1$ suggests that $Q > 3(1+\omega_m)/\lambda - \lambda$. However, since ω_{eff} depends on Q, λ and ω_m and the requirement for acceleration is $\omega_{eff} < -1/3$, these points can generate acceleration if $Q > \lambda(1+3\omega_m)/2$ is satisfied (see Figure 1).

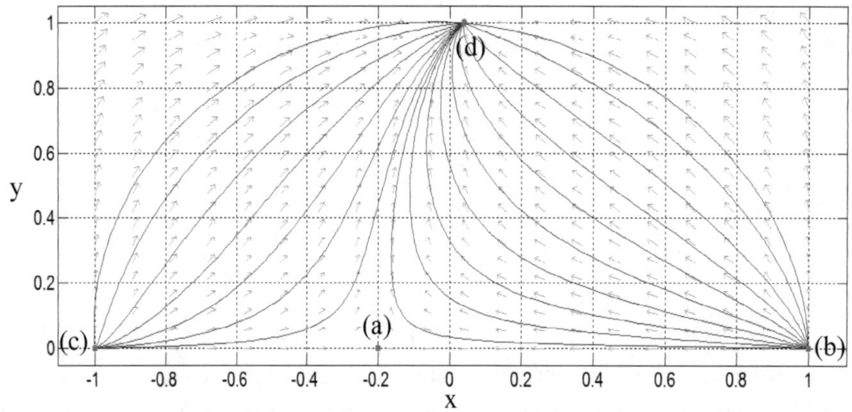

Figure 1. Ordinary field phase diagram for $\lambda = 0{,}1$ and $Q = 0{,}245$. Acceleration solution is given by point (d) (stable node), and matter dominated era by point (a) (saddle point).

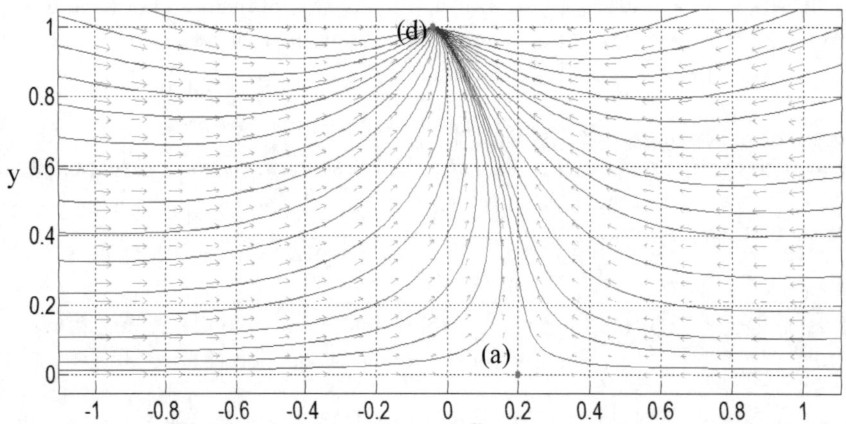

Figure 2. Phantom field phase diagram for $\lambda = 0{,}1$ and $Q = 0{,}245$. Acceleration solution is given by point (d) (stable node) and matter dominated era by point (a) (saddle point).

2.3.2. Phantom scalar field ($\varepsilon = -1$). Point (a) is a kinetic dominated solution ($y = 0$). The condition $\varepsilon = -1$ in the second term keeps this term always negative. For matter with $\omega_m = 0$ we get $-2Q^2/3 < -1/3$, meaning that acceleration can be reached if $Q < \sqrt{1/2}$ is satisfied. However, if we are considering energy fraction, acceleration will not occur for $0 \leq \Omega_\phi \leq 1$. Points (b) & (e) are irrelevant because they are not real solutions. Points (d) & (e) are scalar field dominated solutions ($\Omega_\phi = 1$). Cosmic acceleration will always be satisfied by these points because λ^2 will always be positive, so that $\lambda^2/3\varepsilon$ will always be negative, and $\omega_{eff} = \omega_\phi < -1/3$. So to speak, this is a special property of the phantom field, i.e. $\omega_\phi = \omega_{eff} \leq -1$. Points (f) & (g) just exist if $2Q(Q+\lambda) > 3(1-\omega_m^2)$ is fulfilled. Cosmic acceleration occurs if $Q > \lambda(1+3\omega_m)/2$ is satisfied (See Figure 2).

Table 1. Critical points and cosmological parameters generated from the models

Point	x	y	Ω_ϕ	ω_ϕ	ω_{eff}
(a)	$-\frac{\sqrt{6}Q}{3\varepsilon(1-\omega_m)}$	0	$\frac{2}{3\varepsilon}\left(\frac{Q}{(1-\omega_m)}\right)^2$	1	$\omega_m + \frac{2}{3\varepsilon}\frac{Q^2}{(1-\omega_m)}$
(b)	$\frac{1}{\sqrt{\varepsilon}}$	0	1	1	1
(c)	$-\frac{1}{\sqrt{\varepsilon}}$	0	1	1	1
(d)	$\frac{\lambda}{\sqrt{6\varepsilon}}$	$\left[\frac{1}{V_0}\left(1-\frac{\lambda^2}{6\varepsilon}\right)\right]^{1/2}$	1	$-1+\frac{\lambda^2}{3\varepsilon}$	$-1+\frac{\lambda^2}{3\varepsilon}$
(e)	$\frac{\lambda}{\sqrt{6\varepsilon}}$	$-\left[\frac{1}{V_0}\left(1-\frac{\lambda^2}{6\varepsilon}\right)\right]^{1/2}$	1	$-1+\frac{\lambda^2}{3\varepsilon}$	$-1+\frac{\lambda^2}{3\varepsilon}$
(f)	$\frac{\sqrt{6}(1+\omega_m)}{2(\lambda+Q)}$	$\left[\frac{2Q(\lambda+Q)+3\varepsilon(1-\omega_m^2)}{2V_0(\lambda+Q)^2}\right]^{1/2}$	$\frac{Q(\lambda+Q)+3\varepsilon(1+\omega_m)}{(\lambda+Q)^2}$	$\frac{-Q(\lambda+Q)+3\varepsilon\omega_m(1+\omega_m)}{Q(\lambda+Q)+3\varepsilon(1+\omega_m)}$	$\frac{\omega_m\lambda-Q}{\lambda+Q}$
(g)	$\frac{\sqrt{6}(1+\omega_m)}{2(\lambda+Q)}$	$-\left[\frac{2Q(\lambda+Q)+3\varepsilon(1-\omega_m^2)}{2V_0(\lambda+Q)^2}\right]^{1/2}$	$\frac{Q(\lambda+Q)+3\varepsilon(1+\omega_m)}{(\lambda+Q)^2}$	$\frac{-Q(\lambda+Q)+3\varepsilon\omega_m(1+\omega_m)}{Q(\lambda+Q)+3\varepsilon(1+\omega_m)}$	$\frac{\omega_m\lambda-Q}{\lambda+Q}$

3. Perturbation dynamics

Matter perturbation can be regarded as a small deviation of Einstein field equation from its background value. In the linear regime perturbed gravity can be written as follows

$$\delta\left[R_{\mu\nu} - \frac{1}{2}g_{\mu\nu}R\right] = \delta\left[8\pi G T_{\mu\nu}\right]. \tag{10}$$

Metric tensor of perturbed universe in a conformal-Newtonian gauge is given by

$$g_{\mu\nu} = a^2\begin{pmatrix} -(1+2\Phi) & 0 \\ 0 & (1-2\Psi)\delta_{ij} \end{pmatrix}. \tag{11}$$

For our necessity, in this section we assume that the universe contains only matter and scalar field, which are both perfect fluids and obey energy momentum tensor $T_{\mu\nu} = (\rho + P)u_\mu u_\nu + Pg_{\mu\nu}$. By solving eq. (10) with eq. (11) and defining $c_s^2 = \dot{P}/\dot{\rho}$ and $\delta = \delta\rho/\rho$, and then expanding eq. (3) up to linear perturbation, we obtained

$$3\mathrm{H}(\Phi' - \mathrm{H}\Phi) - k^2\Phi = 4\pi G a^2 \delta\rho,$$
$$k^2(\mathrm{H}\Phi' + \Phi) = 4\pi G a^{-2}(1+\omega)\rho\theta,$$
$$\Phi'' + 3\mathrm{H}\Phi' + (2\mathrm{H}' - \mathrm{H}^2)\Phi = 4\pi G a^2 c_s^2 \delta\rho,$$

$$\delta' = 3Q\phi'\left(\omega - c_s^2\right)\delta + 3\mathrm{H}\left(\omega - c_s^2\right)\delta - (1+\omega)(\theta + 3\Phi') + Q(1-3\omega)\left(\hat{\phi}\right)', \tag{12}$$

where $\hat{\phi} \equiv \delta\phi$. Here we have transformed the equations into Fourier space, where k denotes the wavenumber of perturbation. Eq. (12) is reduced to uncoupled case for $Q = 0$. From the last term of the above equations, we see that the existence of coupling Q connects scalar field perturbation to matter perturbation. As a result, the deviation of scalar field will affect the evolution of the matter perturbation. The explicit form of $\hat{\phi}$ can be obtained by perturbing eq. (2), and for sub-horizon scale it yields

$$\hat{\phi} = -3\lambda_k^2 Q\Omega_m\left(1-\Omega_\phi\right)\delta_m,$$

where $\lambda_k^2 \equiv aH/k$. In $N \equiv \ln a$ coordinate and applying $\omega \approx c_s^2 \ll 1$ eq. (12) takes a simpler form

$$\frac{d\delta_m}{dN} = -\hat{\theta}_m - 3\frac{d\Phi}{dN} + Q\frac{d\hat{\phi}}{dN}. \tag{13}$$

To solve eq. (13) we need an equation of velocity perturbation, and it has been given by [16]. In our notation it is written as

$$\frac{d\hat{\theta}_m}{dN} = -\left(2 + \frac{1}{H}\frac{dH}{dN} + \sqrt{6}Qx\right)\hat{\theta}_m + \frac{1}{\lambda_k^2}\left(\Phi + Q\hat{\phi}\right). \tag{14}$$

From eq. (13) and (14) we finally obtained the perturbation equation

$$\frac{d^2\delta_m}{dN^2} + \left(\frac{1}{2} - \frac{3}{2}\omega_{eff} + \sqrt{6}Qx\right)\frac{d\delta_m}{dN} - \frac{3}{2}\left(1-\Omega_\phi\right)\left\{1 + 2Q^2\left(1-\Omega_\phi\right)\right\}\delta_m = 0. \tag{15}$$

Eq. (15) is a second-order non-linear differential equation, which is difficult to solve analytically because of its complexity in terms of time. However, we can still approximate its solutions by recalling our results in the previous section.

3.1. *Matter dominated era*

This regime is associated with point (a) in Table 1. For matter with $\omega_m = 0$ the parameters are: $x = -\sqrt{6}Q/3\varepsilon$, $\Omega_\phi = 2Q^2/3\varepsilon$, and $\omega_{eff} = 2Q^2/3\varepsilon$. The solution is $\delta(a) = C_1 a^{\gamma_1} + C_2 a^{\gamma_2}$, where

$$\gamma_{1,2} = \frac{1}{\varepsilon}\left[\frac{3Q^2}{2} - \frac{\varepsilon}{4} \pm \frac{1}{12}\sqrt{192Q^6 - 576\varepsilon Q^4 + 432\varepsilon^2 Q^2 + 3324Q^4 - 252\varepsilon Q^2 + 225\varepsilon^2}\right].$$

For the ordinary field ($\varepsilon = +1$),$Q \neq 0$ implies $\gamma > 1$, which means that for this mode structures grow faster than in uncoupled cases, since ω_{eff} is not exactly equals to zero, but there exists a small amount of energy which is transformed into matter via the coupling Q. In contrast, phantom field ($\varepsilon = -1$) makes structures grow slower than in uncoupled cases, due to its negative energy that inverts the flow of the energy current.

3.2. *Matter scaling era*

This regime is associated with point (f). For matter scaling with $\omega_m = 0$, it is easily obtained that $Qx = -\sqrt{6}\omega_{eff}/2$. The solution is similar to that in subsection 3.1, with

$$\gamma_{1,2} = \frac{1}{4}\left[9\omega_{eff} - 1 \pm \sqrt{96Q^2\Omega_\phi^2 - 144Q^2\Omega_\phi + 48Q^2 + 81\omega_{eff}^2 - 48\Omega_\phi - 18\omega_{eff} + 25}\right]. \tag{16}$$

Growth of matter perturbation in this regime is determined by three quantities: Q, ω_{eff} and Ω_ϕ. Eq. (16) shows that $\gamma_1 > 1$ for $Q \neq 0$. However, we do not expect an excessive value of γ, because it has not been expected by observation of Integrated Sach-Wolfe (ISW) effect [13]. Phantom field corresponds to the condition $\Omega_\phi + \omega_{eff} < 0$, where the solutions are negative and complex with negative real parts. The complex solutions with negative real part cause perturbation to decays and oscillates harmonically. In [17] this phenomenon is called phantom damping.

3.3. Scalar field dominated era

This regime is associated with point (d) in Table 1. The solution is similar to that in previous subsection, with

$$\gamma_{1,2} = \frac{1}{2}\left[-5 + \frac{3\lambda^2}{2\varepsilon} \pm \left(5 - \frac{3\lambda^2}{2\varepsilon}\right)\right]. \tag{17}$$

γ_1 will always be zero for all λ, which leads to constant mode solutions. Whereas, γ_2 depends only to λ, i.e. $\gamma_2 = -5 + 3\lambda^2/2\varepsilon$. As explained in the previous section, point (d) will trigger acceleration solutions if $\lambda < \sqrt{2}$ is satisfied, therefore it leads the universe towards the condition $\gamma_2 < -5 + 6/2\varepsilon$. In the eq. (17) we see that at point (d) the coupling effect is vanished, since the matter fraction in the universe is nearly zero. An ordinary field ($\varepsilon = +1$) implies $\gamma_2 < -2$, meaning that density contrast will decay in time, whereas phantom field implies $\gamma_2 < -8$. The decay of density contrasts following phantom scenario will occur faster than that of ordinary field does. This indicates that background dynamics in phantom model expands more explosively compared to that in ordinary models. This result also suggests that big rip will be the relevant fate of our universe if phantom field is the dark energy we are searching for.

4. Conclusions

We have shown that a universe containing a coupled scalar field would have different dynamical evolution and effects to structures growth compared to uncoupled cases. By investigating the dynamics through stability analysis of its critical points we have found the critical points of the universe that related to physical regimes such as radiation dominated and its scaling era, matter dominated and its scaling era, and finally the solutions will be caught by stable solutions, i.e. scalar field dominated solutions. During matter dominated era, following ordinary field scenario structures grow faster than in phantom field and uncoupled cases, whereas following phantom field scenario structures grow slower than both uncoupled and ordinary cases. Far in the future, the decay of structures following phantom scenario occurs drastically faster compared to that of ordinary field scenarios. This is related to background dynamics, which shows that big rip will be the relevant fate of our universe.

References

[1] Perlmutter S, et. al. 1999 Measurement of Ω and Λ from 42 High–Redshift Supernovae *Astrophys. J* **517** 565
[2] Riess A G et. al. 1998 Observational Evidence from Supernovae for an Accelerating Universe and A Cosmological Constant *Astron. J* **116** 1009
[3] Weinberg S 1989 The Cosmological Constant Problem *Rev. Mod. Phys.* **61** 1
[4] Wetterich C 1995 The cosmon model for a asymtotically vanishing time dependent cosmological constant *Astron. Astrophys.* **301** 321
[5] Ratra B and Peebles P J E 1988Cosmological consequences of a rolling homogeneous scalar field *Phys. Rev.* D **37** 3406
[6] Caldwell R R 2000 An Introduction to Quintessence *Brazilian Jounal of Physics* **30** 2
[7] Mulki F A M and H R T Wulandari 2017 Growth of Matter Perturbation in Quintessence Cosmology *American Institute of Physics* **1801** 030002
[8] Amendola L 1999 Scaling solution in general non–minimal coupling theories *Phys. Rev.* D **60** 043501
[9] Amendola L 2000 Coupled Quintessence *Phys. Rev.* D **62** 043511
[10] Bertolami O 2000 Nonminimal coupling and quintessence *Phys. Rev.* D **61** 064007
[11] Pettorino V 2008 Coupled and Extended Quintessence: theoretical differences and structure formation *arXiv: 0802.1086v2*
[12] Gumjudpai B, Naskar and Tsujikawa S 2008 Coupled Dark Energy: Towards a general description of the dynamics *arXiv:hep-th/0502191v2*

Journal of Physics: Conference Series **1354** (2019) 012007 doi:10.1088/1742-6596/1354/1/012007

[13] Copeland E J, Sami M and Tsujikawa S 2006 Dynamics of Dark Energy *Int. J. Mod. Phys.* D **15** 1753
[14] Planck Collaboration 2016 Planck 2015 results. XIV. Dark energy and modified gravity *A & A* **A** 14 594
[15] Piazza FandTsujikawa S 2004 *JCAP* **0407** 004
[16] Amendola L 2004 Phantom Energy Mediates a Long-Range Repulsive Forces *Phys. Rev. Lett.* **93** 181102
[17] Amendola L, Tsujikawa S and Sami M 2006 Phantom damping of matter perturbation. *Phys. Lett.* B **632** 155
[18] Copeland E J, A. R. Liddle and Wands D 1998 *Phys. Rev.* D **57** 4686

The effective dark energy in brane universe

B J Bansawang, M N Gazali Yunus, Azwar Sutiono and Tasrief Surungan

Theoretical and Computational Physics Laboratory, Department of Physics, Hasanuddin University, Makassar, South Sulawesi 90245, Indonesia

E-mail: yunusmng13h@student.unhas.ac.id, tasrief@unhas.ac.id

Abstract. We study the dynamical system of braneworld cosmology with the effective cosmological constant as the effective dark energy on brane universe. Firstly we introduce several scenarios of braneworld model then derive the effective EFE on the brane which reduced at low energy limit. Next, we discuss the FRW cosmological model in braneworld scenario considered in static and non-closed system. In the dynamical study of cosmology, we assume the equation of state of dark energy varies with time and can be parameterized as a function of scale factor of the universe. Using the dynamical system approach, we describe the evolution of brane universe with non-vanishing NCC/ BCC term. In this study, we investigate the effects of the presence of NCC/ BCC perturbations in brane universe. Finally, we discuss the NCC/ BCC perturbations in the late-time cosmic acceleration when the dark energy contribution is $\Omega_\Lambda \simeq 0.72$.

1. Introduction

The study of modified Einstein field equation (EFE) in general relativity is recently an active research topic in theoretical physics. One type of modification is braneworld scenario using the extra-dimensional approach. The braneworld model was motivated by M-theory, first proposed by Horava and Witten who showed that the $E_8 \times E_8$ heterotic string theory is equivalent to the realization of M-theory on the eleven-dimensional (11D) orbifold $\mathbf{R}^{10} \times \mathbf{S}^1/Z_2$ [1]. This theory can be compactified into $\mathbf{R}^4 \times \mathbf{S}^1/Z_2 \times$ [Calabi-Yau]^6. As the tiny 6D Calabi-Yau extra space was smaller than other dimensions, the theory can be reduced into the 5D spacetime $\mathbf{R}^4 \times \mathbf{S}^1/Z_2$, where 4D spacetime $\mathbf{R}^4$ was set as boundary layers [2]. In the braneworld scenario based on the Horava-Witten theory, the 4D spacetime was described as a domain wall called a brane and the 5D full spacetime as the bulk spacetime. The standard model particles are confined on a brane, whereas gravitons propagate in the bulk spacetime [3]. In this case, the matter fields are confined in the 4D spacetime while gravity acts in the full spacetime.

In addition to the formulation by Horava-Witten, the braneworld scenario was also formulated by Randall and Sundrum (RS) in the form of 5D warped spacetime [4,5], and by Dvali, Gabadadze, and Porrati (DGP) with 5D Minkowski spacetime [6]. There are two models in the RS formulation, the first RS model (RS1 model) explains that two identical branes are embedded in a 5D Anti-de-Sitter (AdS) bulk spacetime with Z_2 symmetry and small extra dimension [4]. In the second RS model (RS2 model), a single brane with positive tension (λ) is embedded in a 5D AdS bulk spacetime of infinite extra dimensions [5]. Whereas, in the DGP model, the 3-brane is embedded in a Minkowski bulk spacetime of infinitely large extra dimensions [6].

Content from this work may be used under the terms of the Creative Commons Attribution 3.0 licence. Any further distribution of this work must maintain attribution to the author(s) and the title of the work, journal citation and DOI.
Published under licence by IOP Publishing Ltd

Derivation of the EFE in braneworld scenario was proposed by Shiromizu, *et al.* [7], who showed that the effective EFE reduced to the conventional EFE for low energy limit. The effective EFE on brane is obtained by projecting the 5D metric and the 5D EFE onto the brane volume. The effective EFE showed a non-closed system by the presence of 5D Weyl correction term as a bulk nonlocal effect on brane. However, the effective cosmological constant and Λ_5 were not discussed in Ref. [7]. In another study, Maartens, *et al.* [8] derived the effective EFE, generalizing the effective field equation in Ref. [7] with an additional correction term. The derived cosmology in braneworld scenario with a perfect fluid model was introduced in Ref. [8]. However, the effective cosmological constant was negligible for the dynamics of braneworld cosmology in Ref. [8].

In this paper, we report our study on the dynamical system of braneworld cosmology, and try to describe the effects of the presence of nonlocal cosmological constant (NCC) or bulk cosmological constant (BCC) perturbations in brane universe. The evolution of brane universe is described without ignoring the NCC/ BCC term, i.e. Λ_5 as the nonlocal effect on the brane. We also discuss the Friedmann-Robertson-Walker (FRW) brane as a non-closed system and static with the effective cosmological constant contributing at low energy limit. In this case, the matter field terms associated with high energy can be negligible, whereas the bulk nonlocal effects on the brane are non-vanishing for the non-closed brane system. We argue that the dark energy is defined by the effective cosmological constant expressed as the total energy density of the vacuum energy localized on brane and the NCC/ BCC perturbations, where Λ_5 could play a very important role, especially in the late-time cosmic acceleration.

The remaining part of the paper is organized as follows: section 2 describes the effective field equation on the brane for low energy limit. The dynamical study of braneworld cosmology and the effects of the NCC/ BCC perturbations on FRW brane are discussed in section 3 and section 4, respectively. Section 5 is devoted to the concluding remarks.

2. The effective EFE

In the braneworld scenario, the 4D spacetime is described by a domain wall, i.e. 3-brane $(\Sigma, q_{\mu\nu})$ embedded in 5D bulk spacetime $(\mathcal{M}, g_{AB})$. The induced metric on Σ in full spacetime is given by $q_{AB} = g_{AB} - n_A n_B$, with n_A being a vector unit normal to Σ and g_{AB} is the 5D metric. The EFE in the bulk spacetime, with negative cosmological constant Λ_5 (see Ref. [7,8]), is given as

$$^{(5)}G_{AB} = -\Lambda_5\; g_{AB} + \kappa_5^2\; {}^{(5)}T_{AB}, \tag{1}$$

where $^{(5)}G_{AB}$ is the 5D Einstein tensor and κ_5^2 is the 5D gravitational constant. $^{(5)}T_{AB}$ is the energy-momentum tensor of the matter fields in the bulk, whereas Λ_5 is the bulk cosmological constant (NCC/ BCC term). Since there are no matter fields other than gravity in the bulk spacetime for braneworld scenario, we can assume that $^{(5)}T_{AB} = 0$, so that the 5D EFE becomes $^{(5)}G_{AB} = -\Lambda_5\; g_{AB}$ in the bulk [8]. Then, by using the 5D metric projections on the brane from Gauss-Codazzi-Mainardi equations [7–10], we have the following effective EFE on the brane:

$$^{(4)}G_{\mu\nu} = -\Lambda_4\; q_{\mu\nu} + \kappa^2\; T_{\mu\nu} + \frac{6}{\lambda}\kappa^2\; \Pi_{\mu\nu} - E_{\mu\nu}, \tag{2}$$

where the extrinsic curvature on Σ is uniquely determined by the Israel junction condition and imposed the Z_2 symmetry with the brane as the fixed point in the full spacetime. Here, $\kappa^2 = 8\pi G_N = \frac{1}{6}\kappa_5^4\lambda$ is the 4D gravitational constant and $T_{\mu\nu}$ being the energy-momentum tensor of the matter fields on Σ. The effective cosmological constant on the brane [8,10] is written as

$$\Lambda_4 \equiv \frac{1}{2}\left(\Lambda_5 + \kappa^2\lambda\right), \tag{3}$$

where λ being the brane tension or the vacuum energy localized on the brane. Based from RS model [4,5] (see also Ref. [9,10]), Λ_5 is defined as

$$\Lambda_5 \equiv -\frac{6}{\ell^2}. \tag{4}$$

In the effective field equation (2) on the brane, there are two additional correction terms which generalize the conventional EFE. The first correction term is:

$$\Pi_{\mu\nu} \equiv \frac{1}{12}TT_{\mu\nu} - \frac{1}{4}T_{\mu\alpha}T^{\alpha}_{\ \nu} + \frac{1}{8}q_{\mu\nu}T_{\alpha\beta}T^{\alpha\beta} - \frac{1}{24}q_{\mu\nu}T^2, \tag{5}$$

which arises from the extrinsic curvature terms and quadratic in $T_{\mu\nu}$. This term could play an important role in the early universe when the matter-energy scale is high. The second correction term is the projected 5D Weyl tensor term,

$$E_{\mu\nu} \equiv {}^{(5)}C^{A}_{\ \ BRS}\ n_A n^R q^B_\mu q^S_\nu, \tag{6}$$

which carries some information about the bulk nonlocal gravitational field on the brane.

By applying the contracted Bianchi identities in Eq. (2) ($D^\mu\ {}^{(4)}G_{\mu\nu} = 0$), we have

$$D^\mu E_{\mu\nu} = \frac{6}{\lambda}\kappa^2 D^\mu \Pi_{\mu\nu}, \tag{7}$$

in which we imposed the conservation equation $D^\nu T_{\mu\nu} = 0$. This equation shows that the $E_{\mu\nu}$ term as the bulk nonlocal gravitational effect on the effective field equation, may strongly be the gravitational field interaction in high energy regime, because its divergence is constrained by the quadratic term relevant with high energy scale of the matter-energy. Following Ref. [7,9], in the low energy limit, the $\Pi_{\mu\nu}$ and the $E_{\mu\nu}$ terms are negligible. Therefore, the effective field equation (2) on the brane can be reduced to the conventional EFE,

$${}^{(4)}G_{\mu\nu} \simeq -\Lambda_4\ q_{\mu\nu} + \kappa^2\ T_{\mu\nu}, \tag{8}$$

for low energy limit. In this study, Λ_4 is still defined by the effective cosmological constant (3), where Λ_5 is non-vanishing since the brane considered as a non-closed system.

3. The dynamics of braneworld cosmology

3.1. Cosmology of FRW brane

As an alternative approach, the FRW cosmological model in braneworld scenario can be determined by using the Gaussian normal coordinates in the general 5D metric with the cosmological symmetries, given by (see Ref. [8,11])

$$ds^2 = -N^2(t,y)dt^2 + A^2(t,y)\gamma_{ij}dx^i dx^j + dy^2, \tag{9}$$

where $n_\mu dx^\mu = dy$ is the Gaussian normal coordinate orthogonal to the brane. In the Gaussian normal coordinates, the brane is considered static (not moving but located at a fixed coordinate), which can be set as $y = 0$ on the brane. Therefore, the metric (9) can describe the cosmology of FRW brane, with $A(t, y = 0) \equiv a(t)$ being the scale factor on the FRW brane whereas $N(t, y = 0) \equiv 1$ if t is set as proper time on Σ. For this condition, the metric (9) can properly represent the conventional FRW metric on the static brane.

In the standard cosmological aspects of braneworld, the FRW brane is always considered as a homogeneous and isotropic, where the matter-energy on the brane described as a perfect fluid,

$$T_{\mu\nu} \equiv q_{\mu\nu}\ T = q_{\mu\nu} \cdot \mathrm{diag}(-\rho,\ P,\ P,\ P), \tag{10}$$

with ρ is the matter-energy density and P is the isotropic pressure. Therefore, the modified Friedmann equation on the brane for Eq. (2) is given by

$$H^2 = \frac{\Lambda_4}{3} + \frac{\kappa^2}{3}\rho\left(1 + \frac{\rho}{2\lambda}\right) - \frac{k}{a^2} + \frac{2\,\mathcal{U}}{\kappa^2\lambda}, \tag{11}$$

where $\mathcal{U}$ is the dark radiation [8] (see also Ref. [12]). In the low energy limit, as $\rho \ll \lambda$, so that $\rho/\lambda \to 0$, the last term on Eq. (11) will also vanish. Hence, the Friedmann equation on the brane at low energy limit can be written as

$$H^2 = \frac{\Lambda_4}{3} + \frac{\kappa^2}{3}\rho - \frac{k}{a^2}. \tag{12}$$

Here, we assume that Λ_4 contributes to dark energy on the brane with the equation of state of dark energy $w_\Lambda \neq -1$ and varies with time. In this case, the evolution of Λ_4 [13] is given by

$$\dot{\Lambda}_4 = -3H\left(1 + w_\Lambda\right)\Lambda_4. \tag{13}$$

In the cosmological dynamics, we have time derivative of the Friedmann equation (12) on the brane at low energy limit as

$$\dot{H} = \frac{k}{a^2} - \left(\frac{1 + w_\Lambda}{2}\right)\Lambda_4 - \frac{\kappa^2}{2}\left(\rho + P\right), \tag{14}$$

where the second term is derived from Eq. (13).

3.2. Dynamical system approach

From Eqs. (12) and (14), the following equations of motion for the flat FRW background ($k = 0$) on the brane are:

$$3H^2 = \Lambda_4 + \kappa^2\rho, \tag{15}$$

$$2\dot{H} = -\left(1 + w_\Lambda\right)\Lambda_4 - \kappa^2\left(\rho + P\right), \tag{16}$$

where the total energy density ρ and the isotropic pressure P in Eqs. (15) and (16) are given by $\rho = \rho_m + \rho_r$ and $P = \rho_r/3$, respectively. Here ρ_m and ρ_r are the non-relativistic matter density and the radiation density, respectively [14]. In order to study cosmological dynamics, we also introduce the following dimensionless variables:

$$x_1 \equiv \frac{\sqrt{\Lambda_4}}{\sqrt{3}H}, \qquad x_2 \equiv \frac{\kappa\sqrt{\rho_r}}{\sqrt{3}H}, \tag{17}$$

and the density parameters for dark energy, radiation, and non-relativistic matter are respectively given by,

$$\Omega_\Lambda \equiv {x_1}^2, \qquad \Omega_r \equiv {x_2}^2, \qquad \Omega_m \equiv \frac{\kappa^2\rho_m}{3H^2} = 1 - {x_1}^2 - {x_2}^2. \tag{18}$$

These satisfy the relation of total density parameters on the brane, $\Omega_\Lambda + \Omega_r + \Omega_m = 1$ according to Eq. (15). Differentiating the variables x_1 and x_2 with $N = \ln a$ together by using the Eqs. (13) and (15), we obtain the following autonomous equations:

$$\frac{\mathrm{d}x_1}{\mathrm{d}N} = -\frac{3}{2}\left(1 + w_\Lambda\right)x_1 - x_1\frac{\dot{H}}{H^2}, \tag{19}$$

$$\frac{\mathrm{d}x_2}{\mathrm{d}N} = -2x_2 - x_2\frac{\dot{H}}{H^2}, \tag{20}$$

where $\dot{H}/H^2$ is given by

$$\frac{\dot{H}}{H^2} = -\frac{1}{2}\left(3 + 3w_\Lambda {x_1}^2 + {x_2}^2\right). \tag{21}$$

Here, we have the total effective equation of state on the brane as

$$w_{\text{eff}} = -1 - \frac{2}{3}\frac{\dot{H}}{H^2} = w_\Lambda {x_1}^2 + \frac{{x_2}^2}{3}. \tag{22}$$

Now, we derive the fixed points of the dynamical system in equations (19)–(20), which satisfy the conditions $\mathrm{d}x_1/\mathrm{d}N = \mathrm{d}x_2/\mathrm{d}N = 0$ and set $x_2 = 0$ for dark energy dominated era at low energy limit. In this case, we study the following fixed points:

(i) for $x_1 = 0$, we obtain the saddle point

$$(x_1, x_2) = (0, 0), \quad \Omega_\Lambda = 0, \quad \Omega_m = 1, \quad w_{\text{eff}} = 0. \tag{23}$$

(ii) Λ_4-dominated point (stable point)

$$(x_1, x_2) = (\pm 1, 0), \quad \Omega_\Lambda = 1, \quad \Omega_m = 0, \quad w_{\text{eff}} = w_\Lambda. \tag{24}$$

(iii) de-Sitter point (by setting $w_\Lambda = -1$)

$$(x_1, x_2) = (\pm 1, 0), \quad \Omega_\Lambda = 1, \quad \Omega_m = 0, \quad w_{\text{eff}} = -1. \tag{25}$$

Point (i) is matter-dominated solution (23), while point (ii) is compatible with dark energy dominated solution $(\Omega_\Lambda, \Omega_M) = (1, 0)$, where the cosmic acceleration is realized when $w_{\text{eff}} < -1/3$ [14,15]. As w_Λ is not specified, then we try to vary the equation of state of Λ_4 for dark energy dominated epoch (ii). The trajectories of phase space solutions for variation of w_Λ are given in Fig. 1.

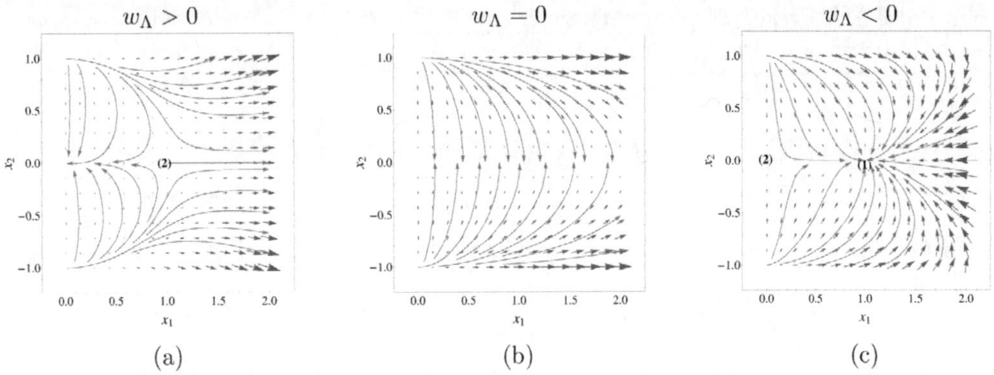

Figure 1. The trajectories of phase space solutions for Eqs. (19)–(20) in the fixed point $(\Omega_\Lambda, \Omega_M) = (1, 0)$ with varying w_Λ. (a) For $w_\Lambda > 0$, critical point (2) at $x_1 = 1$, $x_2 = 0$ is saddle node, (b) for $w_\Lambda = 0$, the solution becomes singular and does not have a fixed point, and (c) for $w_\Lambda < 0$, critical point (1) at $x_1 = 1$, $x_2 = 0$ is stable node, while critical point (2) at $x_1 = 0$, $x_2 = 0$ is saddle node.

Fig. 1 shows that the equation of state of Λ_4 must be smaller than 0 ($w_\Lambda < 0$, $w_\Lambda \in \mathbb{R}^-$) for the current universe, i.e. dark energy dominated epoch (see Fig. 1(c)). In this case, we set w_Λ to satisfy the observational constraints.

Based on the combined data analysis of SN Ia, CMB, and BAO, using the WMAP [14,15], the equation of state for dark energy is given by the bound $-1.098 < w_\Lambda < -0.841$ at the 95% confidence level with dark energy contributing about $\Omega_\Lambda^{(0)} = 0.726 \pm 0.015$ (WMAP 5-year constraint). If $w_\Lambda = -1$ (like in the ΛCDM model), the dark energy density parameter from the CMB shift parameter is constrained to be $0.72 < \Omega_\Lambda^{(0)} < 0.77$ while values of $\Omega_\Lambda^{(0)}$ become smaller for increasing w_Λ [14]. Therefore, we can set the equation of state for dark energy constrained to be $-4/3 < w_\Lambda < -1/3$, for the late-time cosmic acceleration ($w_\Lambda < -1/3$) and the contribution of dark energy compatible with observational data constraints. The trajectories of phase space solutions from the stability point for this case are shown in Fig. 2.

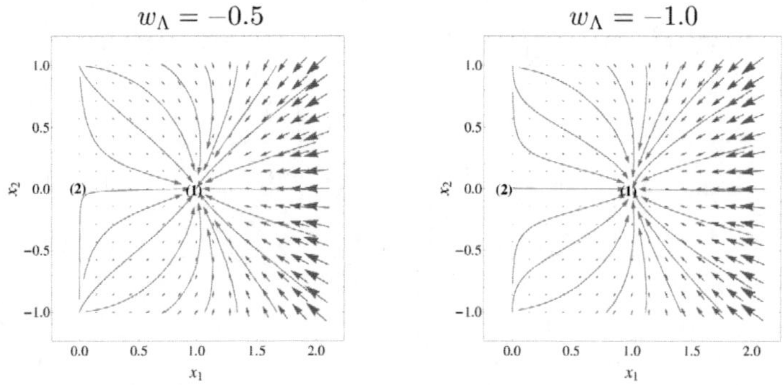

Figure 2. The trajectories of phase space solutions for Eqs. (19)–(20) in the fixed point $(\Omega_\Lambda, \Omega_M) = (1, 0)$ with $-4/3 < w_\Lambda < -1/3$, where critical point (1) at $x_1 = 1$, $x_2 = 0$ is stable node, while critical point (2) at $x_1 = 0$, $x_2 = 0$ is saddle node.

Since the equation of state of Λ_4 varies with time, one needs to parameterize w_Λ to describe its evolution. In this case, we parameterize w_Λ as a function of scale factor of the universe, a [16]. The evolution of w_Λ can be determined by using Eq. (27), with parameterization written in the form (see Ref. [14,16]):

$$w_\Lambda(a) = \sum_{n=0} w_n X_n(a), \; n \leq 1, \tag{26}$$

where the function of expansion is given by $X_n(a) \equiv \left(1 - \frac{a_0}{a}\right)^n$. Then Eq. (26) can be expanded as

$$w_\Lambda(a) = w_0 + w_1 \left(1 - \frac{a_0}{a}\right), \; -4/3 < w_\Lambda(a) < -1/3. \tag{27}$$

For $a = a_0 = 1$, we have $w_\Lambda(a) \to w_0$ and $w_\Lambda(a) = w_0 + w_1$ for $a \to \infty$. Therefore, we found that $w_0 \lesssim -1/3$ and $w_0 + w_1 \gtrsim -4/3$. In Fig. 3, we plot the evolution of Ω_Λ and w_{eff} versus N for each variation of w_Λ, with $w_\Lambda < 0$ and it is expressed as a constant, including for w_Λ as a function of scale factor of the universe (27). For $w_\Lambda = -0.2$, we found the evolution of Ω_Λ slowly increasing and w_{eff} is too large for the cosmic acceleration today. Otherwise, for $w_\Lambda = -2.0$, we found the evolution of Ω_Λ rapidly increasing, while w_{eff} is very small and incompatible with the observational data constraints.

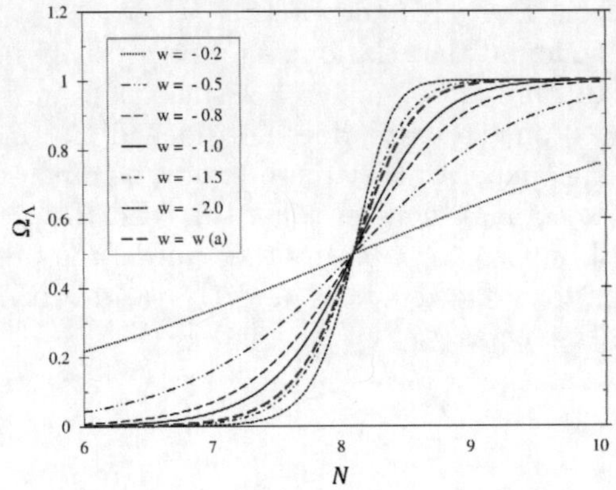

(a)

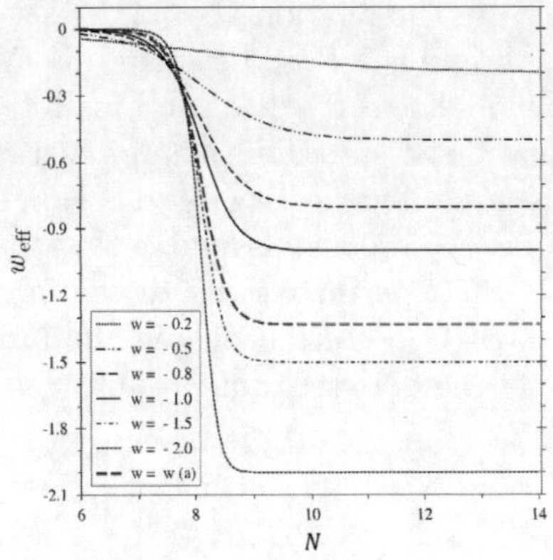

(b)

Figure 3. The evolution of: (a) Ω_Λ, and (b) w_{eff} from Eq. (22), for $w_\Lambda = -0.2; -0.5; -0.8; -1.0; -1.5;$ and -2.0, including for $w_\Lambda \equiv w_\Lambda(a) \simeq -4/3 + a_0 e^{-N}$ which represent the equation (27) with $N = \ln a$ and $a_0 = 1$.

4. The NCC/ BCC perturbations

The NCC/ BCC term is a nonlocal effect on the brane, which could play an important role at low energy limit, especially in the late-time cosmic acceleration. If the NCC/ BCC perturbations are non-vanishing on the brane, we have the total effective vacuum energy density called the effective dark energy given by

$$\rho_\Lambda \equiv \rho_\ell + \rho_\lambda, \tag{28}$$

follows from Eqs. (3) and (4), which satisfy the energy density of 4D cosmological constant, i.e. $\rho_\Lambda = \Lambda_4/\kappa^2$. Here ρ_ℓ and ρ_λ are the NCC/ BCC energy density and the vacuum energy density localized on the brane, respectively, which given by

$$^{(5)}\rho_\Lambda = \frac{\Lambda_5}{2\kappa^2} = -\frac{3}{\kappa^2\ell^2} \equiv \rho_\ell, \tag{29}$$

and

$$\rho_\lambda \equiv \frac{\lambda}{2}. \tag{30}$$

Here, we obtain the total density parameter of the effective dark energy on the brane, follows from Eqs. (28)–(30),

$$\Omega_\Lambda = \frac{\Lambda_4}{3H^2} \equiv -\frac{1}{H^2\ell^2} + \frac{\kappa^2\lambda}{6H^2}. \tag{31}$$

From Eq. (31), we introduce a new additional variable:

$$x_3 \equiv \frac{\kappa\sqrt{\rho_\lambda}}{\sqrt{3}H} = \frac{\kappa\sqrt{\lambda}}{\sqrt{6}H}, \tag{32}$$

with the vacuum energy density parameter localized on the brane and the energy density parameter for the NCC/ BCC perturbations are respectively given as

$$\Omega_\lambda \equiv {x_3}^2, \quad \Omega_\ell \equiv \frac{\Lambda_5}{6H^2} = -(H\ell)^{-2} = {x_1}^2 - {x_3}^2, \tag{33}$$

which satisfy the relation $\Omega_\Lambda = \Omega_\ell + \Omega_\lambda = {x_1}^2$ from Eqs. (18) and (31).

If the brane is a closed system, i.e. $\Lambda_5 = 0$, we found the relation: $\Lambda_4 = \kappa^2 \rho_\lambda$ satisfying the ΛCDM model (clearly discussed in Ref. [17]). In that case, $\dot{\Lambda}_4 = 0 = \dot{\lambda}$ since $w_\Lambda = w_\lambda = -1$ (the equation of state for vacuum energy localized on the brane does not depend on time and has constant value, i.e. $w_\lambda = -1$). However, in this paper, we set the brane to be a non-closed system, i.e. $\Lambda_5 \neq 0$ and assuming that the evolution of Λ_5 is non-zero. For this case, the presence of NCC/ BCC perturbations on the brane could impose $\dot{\Lambda}_4 \neq 0$. In other words, the presence of NCC/ BCC perturbations on the brane leads to $w_\Lambda \neq w_\lambda = -1$ and w_Λ varies with time. Therefore, the derivative of variable x_3 with N is given as

$$\frac{\mathrm{d}x_3}{\mathrm{d}N} = \frac{x_3}{2}\left(3 - 3{x_3}^2 + {x_2}^2\right), \tag{34}$$

with initial condition $x_3 = x_1$, and by setting $w_\lambda = -1$ and $\dot{\lambda} = 0$. Completely, the effect of the NCC/ BCC perturbations on the brane is described in the plot of evolution of the universe components given in Fig. 4.

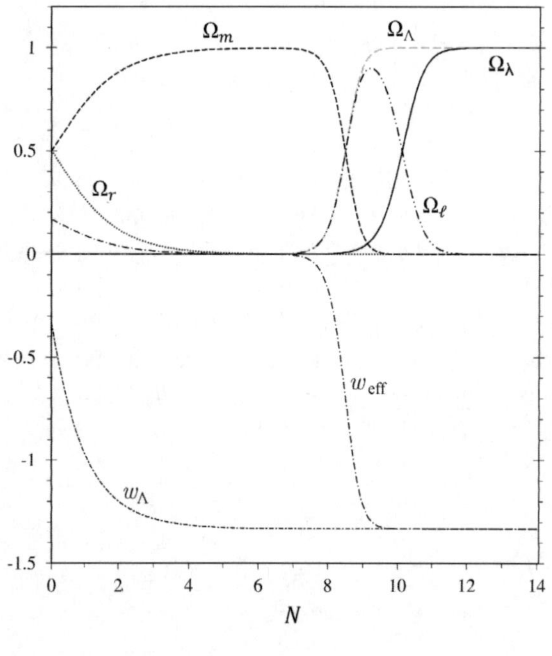

Figure 4. The evolution of Ω_r, Ω_m, Ω_ℓ, and Ω_λ, with $\Omega_\Lambda = \Omega_\ell + \Omega_\lambda$ (the grey dashed-line curve), together with the evolution of $w_\Lambda \equiv w_\Lambda(a)$ and w_{eff}. The initial conditions are given by $x_1 = x_3 = 0.5 \times 10^{-6}$ and $x_2 = 0.707$ at $N = 0$.

It is shown in Fig. 4, the evolution of Ω_r, Ω_m, Ω_Λ, Ω_ℓ, Ω_λ, w_Λ, and w_{eff} versus N with $\Omega_\Lambda = \Omega_\ell + \Omega_\lambda$ is the density parameter of the effective dark energy on the brane, follows from Eq. (31). Starting from the matter-radiation equality epoch, we found the solutions first dwell around the matter-dominated epoch with $w_{\text{eff}} \simeq 0$ and finally approach the dark energy dominated era with $w_{\text{eff}} \simeq w_\Lambda$. Fig. 4 shows that the energy density of the NCC/ BCC term grows with the increase in the density parameter of the effective dark energy Ω_Λ on the brane, but it begins to decrease after the increase of the vacuum energy localized on the brane, Ω_λ. The evolution of Ω_ℓ also has a dynamical solution similar to the evolution of the

Gauss-Bonnet (GB) term which is given in Ref. [14,18]. The GB term has an important role in the accelerated expansion scenario which occurs at the present epoch. Now, if we consider the present universe, based on the observational data constraints [14], the cosmic expansion is accelerated when $w_{\rm eff} < -1/3$, with $\Omega_M \approx 0.3$ and $\Omega_\Lambda \approx 0.7$. From Fig. 4, at about $\Omega_m \simeq 0.28$ and $\Omega_r \simeq 4.2 \times 10^{-5}$, we obtain:

$$\begin{aligned} \Omega_\Lambda &\simeq 0.7203, \\ \Omega_\ell &\simeq 0.7036, \\ \Omega_\lambda &\simeq 0.0167, \\ w_{\rm eff} &\simeq -0.96, \end{aligned} \tag{35}$$

for $N \simeq 8.7$. In Eq. (35), we found the values of the density parameter for the effective dark energy, Ω_Λ and $w_{\rm eff}$ corresponding to the combined data analysis using WMAP, SN Ia, CMB, and BAO which are given in Ref. [14,15]. Based on the energy contribution rate, we also found that the energy contribution of the NCC/ BCC term denoted as Ω_ℓ, is more dominant than the contribution of the vacuum energy localized on the brane. Therefore, Λ_5 or the NCC/ BCC term may strongly be the source for dark energy on the brane, which causes the accelerating cosmic expansion at the present epoch.

5. Summary and conclusion

We have presented the dynamical system of braneworld cosmology with the effective cosmological constant as the effective dark energy on the brane. In the dynamical study, we assume that the equation of state of dark energy varies with time ($w_\Lambda \neq -1$). We found the evolution of Ω_Λ slowly increasing for $w_\Lambda = -0.2$ and $w_{\rm eff}$ is too large for the cosmic acceleration today, while $w_{\rm eff}$ is very small for $w_\Lambda = -2.0$. Therefore, for the observational constraints, $w_\Lambda \equiv w_\Lambda(a)$ is the most effective conditions to describing the late-time cosmic acceleration today as seen in Fig. 3. Furthermore, in this FRW cosmological model, we consider the static brane and the non-closed brane system.

In the low energy limit, we found that the present cosmic acceleration occurs at the dark energy dominated epoch, with the NCC/ BCC perturbations energy contributes about 70%, whereas the vacuum energy localized on the brane is only about 2%, based on Eq. (35). We also found that the total effective equation of state, i.e. $w_{\rm eff} \simeq -0.96$, which satisfies the observational data constraints for the present universe [14,15]. Therefore, dark energy which is observed today, may be due to the NCC/ BCC perturbations. Moreover, since the NCC/ BCC perturbations is non-vanishing on the brane, the energy contribution of Λ_5 can strongly be the source for dark energy on brane universe. Hence, the presence of the NCC/ BCC perturbations play a very important role for evolution of the universe today.

Acknowledgments

This work was supported by BMIS Research Grant of Hasanuddin University, FY 2018. MN GY is grateful to Kemenristekdikti for the support via PKM Grant Programs, FY 2018.

References

[1] Horava P and Witten E 1996 *Nucl. Phys.* B **460** 506
Horava P and Witten E 1996 *Nucl. Phys.* B **475** 94
[2] Lukas A, Ovrut B A, Stelle K S and Waldram D 1999 *Phys. Rev.* D **59** 086001
[3] Arkani-Hamed N, Dimopoulos S and Dvali G 1998 *Phys. Lett.* B **429** 263
Arkani-Hamed N, Dimopoulos S and Dvali G 1999 *Phys. Rev.* D **59** 086004
[4] Randall L and Sundrum R 1999 *Phys. Rev. Lett.* **83**(17) 3370
[5] Randall L and Sundrum R 1999 *Phys. Rev. Lett.* **83**(23) 4690
[6] Dvali G, Gabadadze G and Porrati M 2000 *Phys. Lett.* B **484** 112
Dvali G, Gabadadze G and Porrati M 2000 *Phys. Lett.* B **485** 208

[7] Shiromizu T, Maeda K and Sasaki M 2000 *Phys. Rev.* D **62** 024012
Sasaki M, Shiromizu T and Maeda K 2000 *Phys. Rev.* D **62** 024008
Maeda K and Wands D 2000 *Phys. Rev.* D **62** 124009
[8] Maartens R 2004 *Living Rev. Rel.* **7**(7) 2004
Maartens R and Koyama K 2010 *Living Rev. Rel.* **13**(5) 2010
[9] Maeda K 2002 *Progress of Theoretical Physics Supplement* **148** 59–90
[10] Durrer R 2005 *AIP Conf. Proc.* **782**(1) 202–240
[11] Mukohyama S 2000 *Phys. Lett.* B **473** 241
Binétruy P, Deffayet C, Ellwanger U and Langlois D 2000 *Phys. Lett.* B **477** 285
[12] Maartens R 2000 *Phys. Rev.* D **62** 084023
Maartens R 2001 *Reference Frame and Gravitomagnetism, Proc. of the XXIII Spanish Rel. Meet.* vol 23 p 93–119
[13] Sorkin R D 2007 *AIP Conf. Proc.* **957** 142–153
[14] Amendola L and Tsujikawa S 2010 *Dark Energy: Theory and Observation* (Cambridge: Cambridge University Press)
[15] Kowalski M *et al.* 2008 *Astrophys. J.* **686** 749
[16] Linder E V 2003 *Phys. Rev. Lett.* **90** 091301
[17] Gazali Yunus M N *et al.* 2018 Dynamical System Approach of Braneworld Cosmology in Low Energy Limit with ΛCDM Model *Prosiding Seminar Nasional Fisika Makassar* 2018 (Makassar: FMIPA UNHAS) p 42–46 (Translate version)
[18] Tsujikawa S and Sami M 2007 String-inspired cosmology: Late time transition from scaling matter era to dark energy universe caused by a Gauss-Bonnet coupling *J. Cosmol. Astropart. Phys.* JCAP01(2007)006

Impact of Non-linear Boson Self-interaction on Dark Stars Properties

A. Rahmansyah, A. Sulaksono

Department of Physics, FMIPA, University of Indonesia, Kampus UI Depok, Indonesia

E-mail: rahmansyahandri2@gmail.com, anto.sulaksono@sci.ui.ac.id

Abstract. We study the properties of dark stars using non-linear boson self-interaction models. These stars are compact objects formed from bosonic dark matter. We focus on the properties of dark stars such as the mass and radius relation, tidal deformation, and moment of inertia. In this work, bosonic dark particle mass is set to be 300 MeV and 400 MeV. The bosonic self-interaction potential models are $|\phi|^4$, cosh-Gordon and Liouville which have the same behavior at low densities but relative difference at high densities. Coupling constants on the corresponding self-interaction potentials are determined by the result of numerical simulations CCDM which requires ratio total cross-section to dark particle mass in the range of $0.1\frac{cm^2}{g} \leq \frac{\sigma}{M} \leq 1\frac{cm^2}{g}$.

1. Introduction

The structure of the universe consists of dark and ordinary sectors. Composition of the universe is 96% dark sector and 4% ordinary sector. Dark sector consists of 29% cold dark matter and 67% dark energy [1]. For cold dark matter, the Λ cold dark matter model is commonly chosen because this model is compatible with the cosmic triangle [2]. However, Λ cold dark matter model has several weaknesses. Weaknesses of Λ cold dark matter model has been discussed in details in Ref. [3]. Solution for the weaknesses of Λ cold dark matter model could be self-interaction cold dark matter [4]. Self-interaction cold dark matter can form compact objects such as condensate dark matter stars [5, 6, 7]. In this work, we study dark stars properties using self-interaction bosonic dark matter within relativistic mean field approximation. Furthermore, we are going to use $|\phi|^4$, cosh-Gordon and Liouville potentials [8], respectively. Bosonic dark particle mass is set to be 300 MeV and 400 MeV [3].

2. Formalism

This section consist of the models and properties of dark stars. The models describe an equation of state of dark stars. Properties of dark stars such as mass and radius relation, tidal deformation, and moment of inertia can be calculated by using equation of state of dark stars.

2.1. The Models

We start from $|\phi|^4$ potential model. First, we write Lagrangian density for $|\phi|^4$ potential model. Lagrangian density for $|\phi|^4$ potential model is

$$\mathcal{L} = \partial_\mu \phi^* \partial^\mu \phi - M^2 \left(\phi^* \phi\right) - \lambda \left(\phi^* \phi\right)^2 , \tag{1}$$

Content from this work may be used under the terms of the Creative Commons Attribution 3.0 licence. Any further distribution of this work must maintain attribution to the author(s) and the title of the work, journal citation and DOI.
Published under licence by IOP Publishing Ltd

which M and λ are dark particle mass and dimensionless coupling constant respectively. λ can be determine by the result of numerical simulations CCDM [7-8] i.e.,

$$0.1\frac{cm^2}{g} \leq \frac{\sigma}{M} \leq 1\frac{cm^2}{g}, \tag{2}$$

while the total cross-section for $|\phi|^4$ potential model is

$$\sigma = \frac{\lambda^2}{64\pi M^2}. \tag{3}$$

Note that because the non-linear contributions of cosh-Gordon and Liouville models are small in low density limit. The total cross section of both models close to that of $|\phi|^4$ model. Therefore, for all models we can use following range of λ coupling:

$$\left(\frac{M}{1MeV}\right)^{\frac{3}{2}} < \frac{|\lambda|}{10^{-3}} < 3\left(\frac{M}{1MeV}\right)^{\frac{3}{2}}. \tag{4}$$

In mean field approximation, for lowest energy $k = 0$ plane wave, we obtain number density of bosons and equation of motion of this model, respectively

$$n = J_0 = 2\omega\left(\phi^*\phi\right), \tag{5}$$

and

$$\omega^3 - M^2\omega - \lambda n\left(\hbar c\right)^3 = 0. \tag{6}$$

From energy-momentum tensor [9]

$$T^{\mu\nu} = \frac{\partial\mathcal{L}}{\partial\left(\partial_\mu\phi_\alpha\right)}\partial^\nu\phi_\alpha - \eta^{\mu\nu}\mathcal{L}, \tag{7}$$

which $\eta^{\mu\nu}$ is Minkowski metric tensor, we obtain energy density as

$$\epsilon = \frac{1}{2}\omega n + \frac{1}{2}\frac{M^2 n}{\omega} + \frac{1}{4}\frac{\lambda n^2\left(\hbar c\right)^3}{\omega^2}. \tag{8}$$

By using thermodynamics relation [10]

$$P = n^2\frac{\partial\left(\frac{\epsilon}{n}\right)}{\partial n}, \tag{9}$$

we obtain pressure as the function of energy density of the dark stars. Solving Eq. (6) using numerical methods with n is varied from 0.1 fm^{-3} until 20 fm^{-3}, we obtain ω value. Inserting n and ω value to energy density and pressure, we obtain energy density and pressure values, respectively. We parametrize the energy density and pressure data by using polynomial function up to tenth-order to obtain the equation of state.

The next potential model is cosh-Gordon. Interaction Lagrangian density for this model is defined as

$$\mathcal{L}_{int} = -\alpha M^2\left[\cosh\left(\beta\sqrt{\phi^*\phi}\right) - 1\right], \tag{10}$$

which α and β are constants, respectively. Expanding Eq. (10) into series, and retain only up to second order, we also obtain similar interaction Lagrangian density as

$$\mathcal{L}_{int} = -\alpha M^2\left[\frac{\beta^2\left(\phi^*\phi\right)}{2} + \frac{\beta^4\left(\phi^*\phi\right)^2}{24} + \ldots\right]. \tag{11}$$

By comparing Eq. (11) and Eq. (1), we can obtain following relation

$$\alpha\beta^2 = 2, \tag{12}$$

$$\beta = \frac{2\sqrt{3\lambda}}{M}, \tag{13}$$

which λ can obtain from Eq. (4). In mean-field approximation, for lowest energy $k = 0$ plane wave, we obtain equation of motion of cosh-Gordon model as

$$\omega^2 - \frac{\alpha M^2 \beta}{2}\sqrt{\frac{2\omega}{n(\hbar c)^3}} \sinh\left(\beta\sqrt{\frac{n(\hbar c)^3}{2\omega}}\right) = 0. \tag{14}$$

From energy-momentum tensor, we obtain energy density

$$\epsilon = \frac{\omega n}{2} + \frac{\alpha M^2}{(\hbar c)^3}\left[\cosh\left(\beta\sqrt{\frac{(\hbar c)^3 n}{2\omega}}\right) - 1\right]. \tag{15}$$

From Eq. (9), we obtain pressure of dark stars. Solving Eq. (14) using numerical methods with n is varied from 0.1 fm^{-3} until 20 fm^{-3}, we also obtain ω value. Inserting n and ω value to energy density and pressure, we obtain energy density and pressure value respectively. From energy density and pressure value we will obtain equation of state in a expansion form similar to that of $|\phi|^4$ model.

The last potential is Liouville model. Interaction Lagrangian density for this model is

$$\mathcal{L}_{int} = -\alpha M^2 \left[e^{\beta^2(\phi^*\phi)} - 1\right], \tag{16}$$

which α and β are constants, respectively. Expanding Eq. (16) into series until second order, we obtain interaction Lagrangian density as

$$\mathcal{L}_{int} = -\alpha M^2 \left[\beta^2(\phi^*\phi) + \frac{\beta^4(\phi^*\phi)^2}{2} + \ldots\right]. \tag{17}$$

By comparing Eq. (17) and Eq. (1), we will obtain following relation

$$\alpha\beta^2 = 1, \tag{18}$$

$$\beta = \frac{\sqrt{2\lambda}}{M}, \tag{19}$$

which λ can also obtain from Eq. (4). In mean-field approximation, for lowest energy $k = 0$ plane wave, we obtain equation of motion of Liouville model as

$$\omega^2 - \alpha M^2 \beta^2 e^{\frac{n\beta^2(\hbar c)^3}{2\omega}} = 0. \tag{20}$$

From energy-momentum tensor, we obtain energy density

$$\epsilon = \frac{\omega n}{2} + \frac{\alpha M^2}{(\hbar c)^3}\left[e^{\frac{\beta^2(\hbar c)^3 n}{2\omega}} - 1\right]. \tag{21}$$

From Eq. (9), we also obtain pressure dark stars. Solving Eq. (20) using numerical methods with n is varied from 0.1 fm^{-3} until 20 fm^{-3}, we obtain ω value. Inserting n and ω value to energy density and pressure, we obtain energy density and pressure value, respectively. From energy density and pressure value we will also obtain equation of state in a expansion form similar to that of $|\phi|^4$ model.

2.2. Properties of Dark Stars

Mass and radius relation of dark stars can be described by Tolman Oppenheimer and Tolkov (TOV) equation [11, 12]

$$\frac{dm}{dr} = 4\pi r^2 \epsilon(r), \tag{22}$$

$$\frac{dP}{dr} = \frac{G\epsilon(r)m(r)}{r^2}\left[1+\frac{P(r)}{\epsilon(r)}\right]\left[1+\frac{4\pi r^3 P(r)}{m(r)}\right]\left[1-\frac{2Gm(r)}{r}\right]^{-1}. \tag{23}$$

In binary system, the moment qudrupole of dark star has relation linear with external potential field [13]

$$Q_{ij} = -\gamma \varepsilon_{ij}, \tag{24}$$

which γ is tidal deformability parameter. γ can be found from the equation below

$$\gamma = \frac{2}{3} k_2 R^5, \tag{25}$$

which k_2 and R are electric tidal Love numbers and radius of dark stars, respectively. The dimensionless tidal deformability parameter can be expressed by equation below

$$\Gamma = \frac{2k_2}{3C^5}, \tag{26}$$

which C is compactness $\left(C = \frac{Gm}{R}\right)$. Before we find k_2, we solving first order differential equation [13, 14]

$$ry' + y^2 + Fy + r^2 Q = 0, \tag{27}$$

$$F = \frac{r - 4\pi r^3\left[\epsilon - p\right]}{r - 2m}, \tag{28}$$

$$Q = \frac{4\pi r\left(5\epsilon + 9p + \frac{\epsilon + p}{\frac{\partial p}{\partial \epsilon}} - \frac{l(l+1)}{4\pi r^2}\right)}{r - 2m} - 4\left[\frac{m + 4\pi r^3 p}{r^2\left(1 - \frac{2m}{r}\right)}\right]^2, \tag{29}$$

with initial boundary condition $y(0) = 2$. Next, for $l = 2$ we can calculate k_2 [13, 15]

$$\begin{aligned} k_2 = \; & \frac{8}{5}(1-2C)^2 C^5 \left[2C(y_2 - 1) - y_2 + 2\right] \\ & [2C\left(4(y_2+1)C^4 + (6y_2 - 4)C^3 + (26 - 22y_2)C^2 + 3(5y_2 - 8)C - 3y_2 + 6\right) - \\ & 3(1-2C)^2\left(2C(y_2 - 1) - y_2 + 2\right)\log\left(\frac{1}{1-2C}\right)]^{-1}. \end{aligned} \tag{30}$$

To calculate moment of inertia, we start with the line element of slowly rotating compact objects [16]

$$ds^2 = -e^{2\nu}dt^2 + e^{2\lambda}dr + r^2\left(d\theta^2 + sin^2\theta d\phi^2\right) - 2\omega(r) r^2 sin^2\theta dt d\phi. \tag{31}$$

which $\omega(r)$ are frame-dragging effect due to rotation. From the definition of $\overline{\omega}(r) \equiv \Omega - \omega(r)$, we solving second order ordinary differential equation below

$$\frac{1}{r^4}\frac{d}{dr}\left[r^4 e^{-\nu}\left(1 - \frac{2Gm(r)}{r}\right)^{\frac{1}{2}}\bar{\omega}\right] + \frac{4}{r^4}\left[\frac{d}{dr}e^{-\nu}\left(1 - \frac{2Gm(r)}{r}\right)^{\frac{1}{2}}\right]\bar{\omega} = 0. \tag{32}$$

After solving equation above, we can obtain the moment of inertia of dark stars from equation below

$$\frac{d\bar{\omega}}{dr} = \frac{6GI\Omega}{R^4}. \tag{33}$$

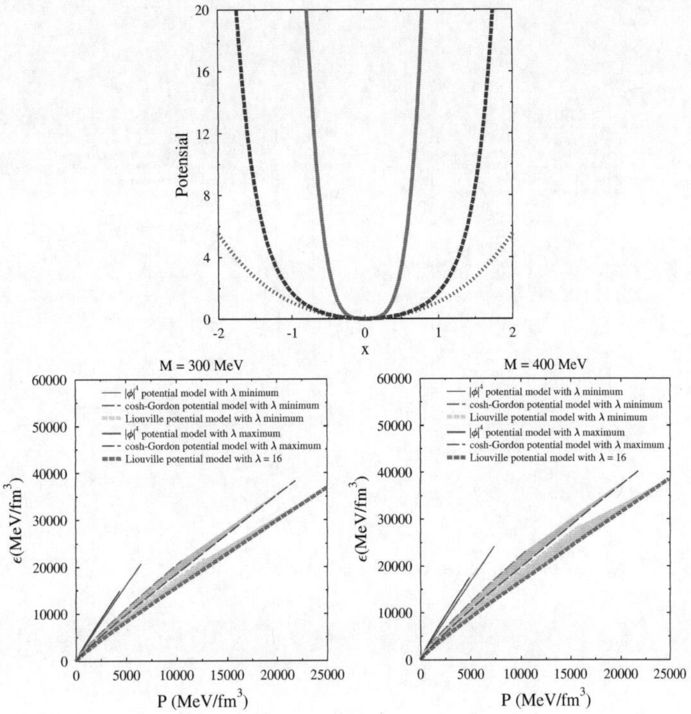

Figure 1. (Top) Difference of potentials, red line is $|\phi|^4$ potential model, blue line is Liouville potential model and green line is cosh-Gordon model. (Bottom) Equation of state with dark particle mass M = 300 MeV (left) and M = 400 MeV (right)

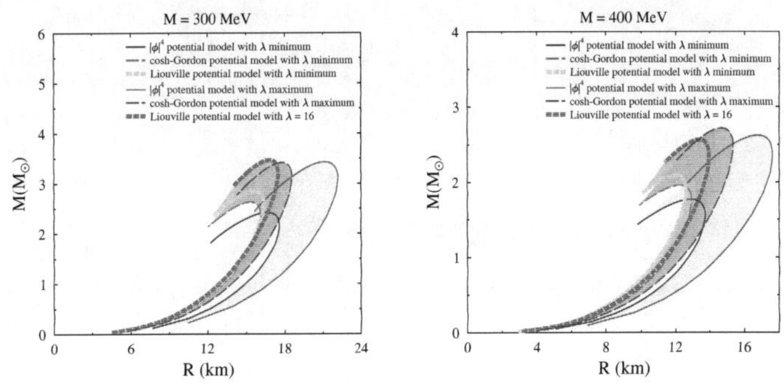

Figure 2. Mass and radius relation dark stars with dark particle mass M = 300 MeV (left) and M = 400 MeV (right)

3. Result and Discussion

It can be seen from each panel in Fig 1 that at low densities, the three models have same potential values. However, at high densities, all models yield significantly different potential

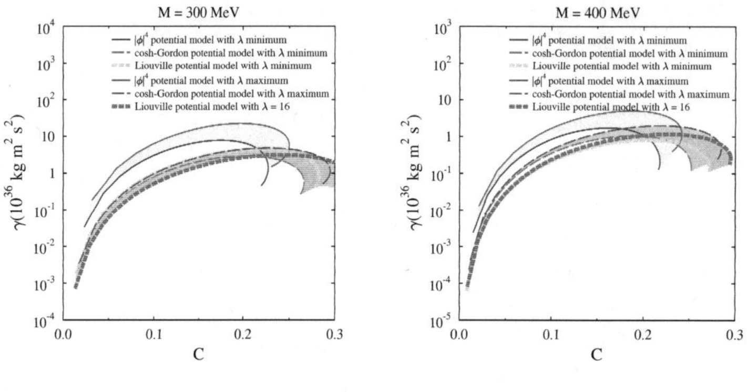

Figure 3. Tidal deformation dark stars with dark particle mass M = 300 MeV (left) and M = 400 MeV (right)

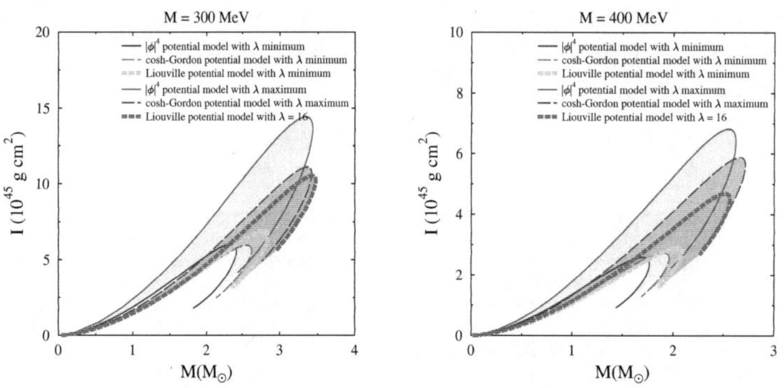

Figure 4. Moment of inertia dark stars with dark particle mass M = 300 MeV (left) and M = 400 MeV (right)

values. Therefore, the differences in dark stars properties come from the stiffness difference in equation of state at high densities. It can be seen obviously from each panel in Fig 2 that at low densities i.e., very low mass and radius region, mass of dark stars with cosh-Gordon and Liouville potential models are close to those of $|\phi|^4$ potential. However, for large mass and radius region, both models show a different behavior than those of with $|\phi|^4$ potential i.e., increasing radius by softening equation of state. It can be seen also all models predict more less the same maximum mass of dark stars. It can be seen also that the maximum mass of dark stars depends significantly on dark particle mass. It can be seen obviously from each panel in Fig 3 the dependency of tidal deformation of the dark stars to the compactness of the corresponding stars. It can seen also that the stiffness difference of the corresponding stars appears more significantly at relative large compactness values. It can be seen also that the tidal deformation parameter of darks stars does not depend on dark particle mass. Fig. 4 show the moment of inertia predicted by all potentials. at low densities i.e., very low mass and radius region, the moment of inertia of dark stars predicted by cosh-Gordon and Liouville potential models are close to those of $|\phi|^4$ potential. However, significant differences in moment of inertia appear at high densities region which are represented by large mass region. It can be seen also that the moment of inertia of darks stars does not depend on dark particle mass.

4. Conclusion

We study the properties of dark stars using non-linear boson self-interaction models using $|\phi|^4$, cosh-Gordon and Liouville bosonic self-interaction potential models. Bosonic dark particle mass is set to be 300 MeV and 400 MeV.The corresponding potentials have the same stiffness behavior at low densities but relative difference at high densities. Coupling constants on the corresponding self-interaction potentials are determined by the result of numerical simulations CCDM which requires ratio total cross-section to dark particle mass in the range of $0.1\frac{cm^2}{g} \leq \frac{\sigma}{M} \leq 1\frac{cm^2}{g}$. We have found that the stiffness difference at high densities are manifested in radii and moment of inertia of large masses dark star, tidal deformation at large masses. While the effect of mass of dark particle appears in maximum mass of dark stars.

ACKNOWLEDGMENT
We are partially supported by the UI's PITTA grant No. 2231/UN2.R3.1/HKP.05.00/2018.

5. References

[1] Freedman W L and Turner M S 2003 Colloquium: Measuring and understanding the universe *Reviews of Modern Physics* **75** 1433
[2] Bahcall N A, Ostriker J P, Perlmutter S, and Steinhardt P J 1999 The cosmic triangle: Assessing the state of the universe *Science* **284** 1481-1488
[3] Maselli A, Pnigouras P, Nielsen N G, Kouvaris C, and Kokkotas K D 2017 Dark stars: Gravitational and electromagnetic observables *Physical Review D* **96** 023005
[4] Spergel D N and Steinhardt P J 2000 Observational evidence for self-interacting cold dark matter *Physical Review Letter* **84** 3760
[5] Li X Y, Harko T, and Cheng K S 2012 Condensate dark matter stars *Journal of Cosmology and Astroparticle Physics* **06** 001
[6] Eby J, Kouvaris C, Nielsen N G, and Wijewardhana L C R 2016 Boson stars from self-interacting dark matter *Journal of High Energy Physics* **02** 028
[7] Rocha M, Peter A H G, Bullock J S, Kaplinghat M, Garrison-Kimmel S, Onorbe J, and Moustaka L A 2013 Cosmological simulations with self-interacting dark matter: I. Constant-density cores and substructure *Monthly Notices of the Royal Astronomical Society* **430** 81
[8] Schunck F E and Torres D F 2000 Boson stars with generic self-interactions *International Journal of Modern Physics D* **09** 601
[9] Glendenning N K 2000 *Compact Stars, Nuclear Physics, Particle Physics, and General Relativity 2nd ed*, editors Appenzeller I, Börner G, Harwit M, Kippenhahn R, Lequeux J, Strittmatter P A, and Trimble V (California: Astronomy and Astrophysics Library) p 154
[10] Glendenning N K 2000 *Compact Stars, Nuclear Physics, Particle Physics, and General Relativity 2nd ed*, editors Appenzeller I, Börner G, Harwit M, Kippenhahn R, Lequeux J, Strittmatter P A, and Trimble V (California: Astronomy and Astrophysics Library) p 175
[11] Tolman R C 1939 Static solutions of Einstein's field equations for spheres of fluid *Physical Review* **55** 364
[12] Oppenheimer J R and Volkoff G M 1939 On massive neutron cores *Physical Review* **55** 374
[13] Hinderer T 2008 Tidal Love numbers of neutron stars *Astrophysical Journal* **677** 1216-1220.
[14] Kumar B, Biswal S K, and Patra S K 2017 Tidal deformability of neutron and hyperon stars within relativistic mean fields equations of state *Physical Review C* **95** 015801
[15] Damour T and Nagar A 2009 Relativistic tidal properties of neutron stars *Physical Review D* **80** 084035
[16] Idrisya A, Owen B J, and Jones D I 2015 R-mode frequencies of slowly rotating relativistic neutron stars with realistic equations of state *Physical Review D* **91** 024001

Journal of Physics: Conference Series **1354** (2019) 012010 doi:10.1088/1742-6596/1354/1/012010

Impact of New Conformal Coupling on Spontaneous Scalarization of Neutron Stars

M. I. Fauzi, H.S. Ramadhan, and A. Sulaksono
Department of Physics, FMIPA, University of Indonesia, Kampus UI Depok, Indonesia
E-mail: muhamad.ihsan41@ui.ac.id, hramad@ui.ac.id, anto.sulaksono@sci.ui.ac.id

Abstract. The scalar-tensor gravity as one of the alternative theories of general relativity has some interesting features. In the case of compact objects it predicts a unique phenomenon which is known as spontaneous scalarization, as shown by Damour and Espasito-Farese [1]. This phenomenon does not appear within the general relativity. In this note, we investigate a new conformal coupling function i.e., $A(\varphi) = 1 - \ln[1 - 0.5\beta(\varphi - \varphi_0)^2]$ and its consequences in neutron stars. In the case of a polytropic equation of state, we have found that the spontaneous scalarization starts to appear when $\beta_c \approx -4.35$. We also found that for the realistic equation of states the β_c depends on the composition and the stiffness of the equation of state.

1. Introduction

A century ago Einstein proposed a modern theory to describe gravitational phenomenon. According to his revolutionary idea, "gravitational field" is the curvature of space-time caused by the existence of massive object(s). This theory is called General Relativity (GR). This is very different concept from Newtonian gravity which assumes gravity as a force between two objects with particular distance. Although very successful, there are several cosmological phenomena that cannot be explained well by GR. This led many people to propose modified gravity theories, as a modification or generalization of GR. One of its candidate is the Scalar Tensor Theory (STT).

The STT was built upon the assumption that gravitational field is carried not only by tensor, as in GR, but also by scalar fields. Observationally, this theory was constrained by the solar system experiment or weak-field gravitation object $\alpha_0^2 < 10^{-3}$, where α is the corresponding coupling constant, which is based on time-delay experiment and light deflection [2]. The simplest, and arguably the most popular, theory in STT is the Brans-Dicke theory [3]. Recently, within the Brans-Dicke framework the STT was shown to have different phenomenon with general relativity as shown by Damour and Espasito-Farese [1]. The phenomenon is known as the "spontaneous scalarization", which is analogous with spontaneous magnetization in magnetic material. In this study, we focus on choosing the new coupling function $A(\varphi)$. Then, the impact of the equation of state (EoS) and mass-radius relation of the neutron star (NS) will also be studied. The NS is assumed to be a static, spherically symmetric, and isotropic matter which possesses ideal fluid characteristic. To determine the value of β when spontaneous scalarization occurs, we use the method proposed by Novak [4].

Content from this work may be used under the terms of the Creative Commons Attribution 3.0 licence. Any further distribution of this work must maintain attribution to the author(s) and the title of the work, journal citation and DOI.
Published under licence by IOP Publishing Ltd

2. Formulation

In the Scalar-Tensor Theory, the action is formulated as

$$S = \frac{1}{16\pi}\int d^4x\sqrt{-g}\Big(\phi R + \frac{\omega}{\phi}g^{\mu\nu}\phi_{,\mu}\phi_{,\nu}\Big) + S_M[\Phi_M; g_{\mu\nu}]. \tag{1}$$

Where S_M is the matter part. The Ricci scalar is coupled directly to the "dilatonic" scalar field ϕ. Matter action contains non-gravitational field variable but it does not contain scalar field. The action in that form is formulated in Jordan frame [1]. The action reduces to general relativity in the limit $\omega \to \infty$ [5].

We could write STT action in similar form with general relativity by using conformal transformation [6] [7]. This transformation makes the new metric coupled with scalar field [8] [9]

$$g_{\mu\nu} \to \tilde{g}_{\mu\nu} = \Omega^2(\varphi)g_{\mu\nu}. \tag{2}$$

This makes $\tilde{g}_{\mu\nu}$ is non-physical metric but could be use in STT representation. Following Damour and Espasito-Farese, we use following transformation

$$d\varphi = \frac{1}{2}\sqrt{2\omega(\phi)+3}\frac{d\phi}{\phi}, \tag{3}$$

$$\Omega^2(\phi) = \phi, \tag{4}$$

$$A(\varphi) = \Omega^{-1}, \tag{5}$$

$$\tilde{\alpha} \equiv \frac{d\ln A(\varphi)}{d\varphi} = \frac{1}{\sqrt{2\omega(\varphi)+3}}. \tag{6}$$

The action now is in Einstein form and could be written

$$S = \frac{1}{16\pi}\int d^4x\sqrt{-\tilde{g}}\Big[\tilde{R} - 2g^{\mu\nu}\varphi_{,\mu}\varphi_{,\nu}\Big] + S_M[\Phi_M; A(\varphi)^2 g_{\mu\nu}]. \tag{7}$$

Where φ is the new scalar field. As could be seen now that the ricci scalar does not coupled with the scalar field but the scalar field is mixed with matter part of the action.

By using variational principle, we could derive field equation of STT in Einstein frame as

$$\tilde{R}_{\mu\nu} = 2\partial_\mu\varphi\partial_\nu\varphi + \frac{8\pi G}{c^4}\Big(\tilde{T}_{\mu\nu} - \frac{1}{2}\tilde{T}\tilde{g}_{\mu\nu}\Big), \tag{8}$$

$$\Box_{\tilde{g}}\varphi = -\frac{4\pi G}{c^4}\alpha(\varphi)\tilde{T}, \tag{9}$$

with $\tilde{T} \equiv 2c\tilde{g}^{-1/2}\delta S_M/\delta\tilde{g}_{\mu\nu}$ indicates the stress-energy tensor in Einstein frame.

In this study, we introduce new coupling function as

$$A(\varphi) = 1 - \ln[1 - 0.5\beta(\varphi-\varphi_0)^2], \tag{10}$$

$$\alpha(\varphi) = \frac{d\ln A(\varphi)}{d\varphi} = \frac{\beta(\varphi-\varphi_0)}{\Big(1-0.5\beta(\varphi-\varphi_0)^2\Big)\Big(1-\ln[1-0.5\beta(\varphi-\varphi_0)^2]\Big)}. \tag{11}$$

In this case, STT theory will reduces to general relativity when $\beta = 0$. The value of $A(\varphi) = 1$ and $\alpha(\varphi) = 0$. This indicates there is no coupling between scalar field with the matter.

By using equation (8) and (9), then assuming the metric has spherical symmetry and static written as

$$ds^2 = -e^{2\Phi(r)}dt^2 + \left(1 - \frac{2m(r)}{r}^{-1} dr^2\right) + r^2 d\Omega^2. \tag{12}$$

The matter inside the neutron star assumed perfect fluid. Stress-energy tensor is expressed as

$$\tilde{T}^{\mu\nu} = (\tilde{\epsilon} + \tilde{p})u^\mu u^\nu + \tilde{p}\tilde{g}^{\mu\nu}. \tag{13}$$

We will get following set of equations

$$\frac{dM}{dr} = 4\pi r^2 A^4 \tilde{\epsilon} + \frac{r}{2}(r - 2M)\psi^2, \tag{14}$$

$$\frac{d\varphi}{dr} = \psi, \tag{15}$$

$$\frac{d\tilde{p}}{dr} = -(\tilde{\epsilon} + \tilde{p})\left[\frac{4\pi r^2 A^4 \tilde{p}}{r - 2M} + \frac{r}{2}\psi^2 + \frac{M}{r(r-2M)} + \alpha\psi\right], \tag{16}$$

$$\frac{d\psi}{dr} = 4\pi \frac{rA^4}{r - 2M}\left[\alpha(\tilde{\epsilon} - 3\tilde{p}) + r\psi(\tilde{\epsilon} - \tilde{p})\right] - \frac{2(r - M)}{r(r - 2M)}\psi. \tag{17}$$

Equation (17) is Tolmann-Oppenheimer-Volkoff (TOV) of STT in Einstein frame. This equation describes the structure of star with spherical symmetry and static.
We use shooting method and Runge-Kutta method to solve equations (14) - (17) numerically [5]. These methods are commonly used to solve differential equation with two boundary conditions. First, we express the system's equation as first order differential equation as shown by equation (14) - (17). Then, we integrate from the centre of the star with following boudary conditions $M(\Delta r) = 0, \varphi(\Delta r) = \varphi_c, \tilde{p}(\Delta r) = 0, \psi(\Delta r) = \frac{4\pi}{3}\Delta r A^4(\varphi_c)(\tilde{\epsilon}_c - 3\tilde{p}_c)$. These boundary conditions are defined at centre of the star except for scalar field which is defined at infinity, $\varphi_0^{max} = 2.4 \times 10^{-3}$. This value follows Damour and Espasito-Farese [1]. The value of scalar field at the centre of star is initially an arbitrary guess. The real value will be obtained when the boundary value at surface is found. In every program's step, the integration moves with distance Δr. This process iterates until the pressure at the surface is zero, $p(R) = 0$. When the boundary is reached, some quantity is calculated by following equations

$$R \equiv r_s, \tag{18}$$

$$\alpha_A \equiv \frac{2\psi_s}{\nu_s'}, \tag{19}$$

$$Q_1 \equiv (1 + \alpha_A^2)^{1/2}, \tag{20}$$

$$Q_2 \equiv \left(1 - \frac{2M_s}{R}\right)^{1/2}, \tag{21}$$

$$\hat{\nu}_s \equiv -\frac{2}{Q_1}\tanh^{-1}\left(\frac{Q_1}{1 + 2(R\nu_s')^{-1}}\right), \tag{22}$$

$$\nu_s' = R\psi_s^2 + \frac{2M_s}{R(R - 2M_s)}, \tag{23}$$

$$\varphi_0 = \varphi_s - \frac{1}{2}\alpha_A\hat{\nu}_s, \tag{24}$$

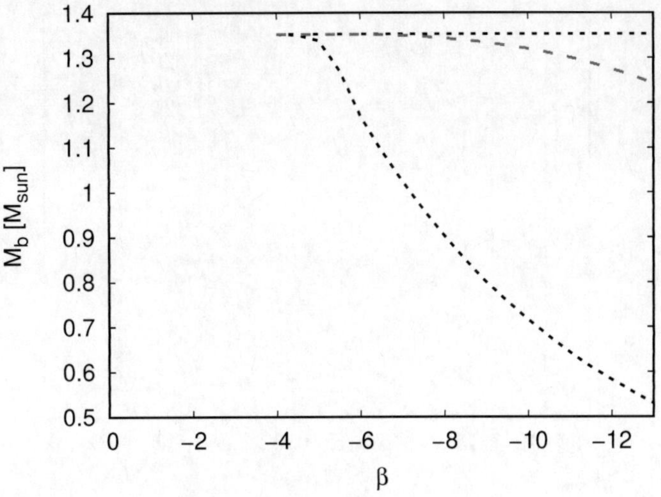

Figure 1. Relation $m_b[M_\odot]$ with β for 1 Central pressure profile $P_c \approx 600\ MeV/fm^3$ for polytropic EoS with $\gamma = 2.34$ and $K = 0.0195$

$$m_A = \frac{1}{2}\nu_s' R^2 Q_2 e^{\frac{1}{2}\hat{\nu}_s}, \tag{25}$$

$$\omega_A = -\alpha_A m_A, \tag{26}$$

$$\bar{m}_A = \bar{M}_s. \tag{27}$$

where m_A is Arnowitt-Deser-Misner (ADM) mass from object A, and ω_A is the scalar charge. Symbol $\bar{m}_A$ is the mass m_A in the Jordan frame.
In this study the EoS used is polytropic EoS [10] with

$$\tilde{p}(\tilde{\rho}) = K\rho_0(\tilde{\rho}/\rho_0)^\gamma \tag{28}$$

where ρ_0 is reference value for rest mass density while K and γ are dimensionless constants. Energy density could be determined by using first thermodynamic law which can be expressed as

$$\tilde{\epsilon}(\tilde{\rho}) = \tilde{\rho} + (\gamma - 1)^{-1}\tilde{p}. \tag{29}$$

We choose polytropic parameters $\gamma = 2.34$, $K = 0.0195$, and $\rho_0 = 1.66 \times 10^{14} g/cm^3$ used commonly in the literatures of nuclear saturation density. We will also use EoS for NS, namely NL3 EoS and G2 EoS. These EoSs are different based on stiffness of matter in NS.

3. Result and Discussion

By using polytropic EoS and we take one central pressure profile in the star. In this case the value $P_c \approx 600 MeV/fm^3$ and varying β we could determine when the spontaneous scalarization occurs, as shown in Figure 1. The value of β starts from $\beta = -4$ to $\beta = -12.5$. In Figure 1, an area inside the curve indicates spontaneous scalarization phenomenon. The horizontal line is the maximum baryonic mass of neutron star when $P_c \approx 600 MeV/fm^3$ which is $M_b \approx 1.35 M_\odot$. This phenomena occurs when $\beta \approx -4.5$.
After we use polytropic equation of state we will use NL3 EoS and G2 EoS. NL3 EoS has softer characteristic than G2 EoS.

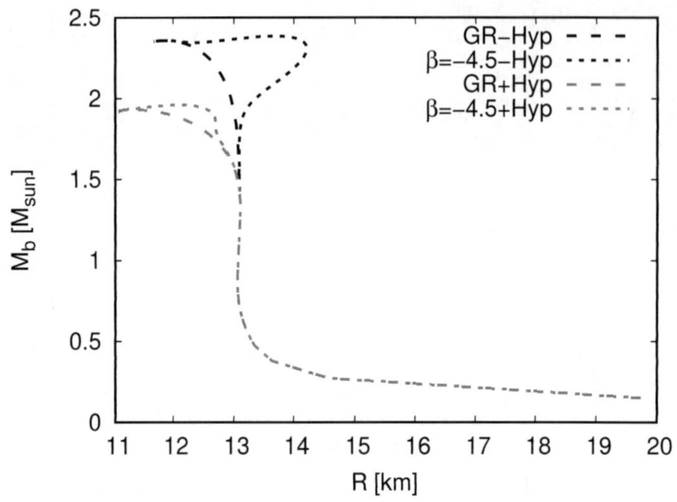

Figure 2. Relation $m_b[M_\odot]$ with $R[km]$ NL3 EoS

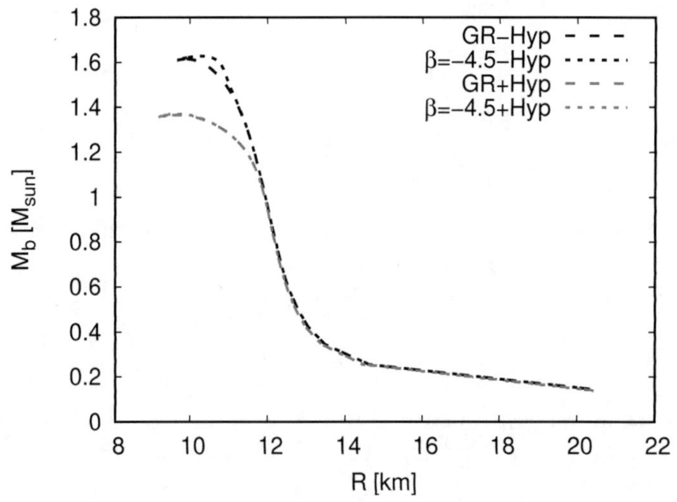

Figure 3. Relation $m_b[M_\odot]$ with $R[km]$ G2 EoS

Mass-radius relation of neutron star shown in Figure 2 and Figure 3. In Figure 2, blue line indicates that of NL3 EoS with hyperon existence. If we compare with GR, there is a deviation or spontaneous scalarization when $\beta = -4.5$. This also happens when we use NL3 EoS without hyperon but with larger scalarization effect in mass-radius relation than with hyperon. From Figure 2, maximum baryonic mass for NL3 EoS without hyperon is larger than maximum baryonic mass predicted by NL3 EoS with hyperon.
In Figure 3, blue line indicates the one G2 EoS with hyperon while the black line for G2 EoS without hyperon. When $\beta \approx -4.5$, scalarization does not appear for G2 EoS with hyperon. This is indicated by the in-line plot of the one of with GR. Then, scalarization appears at $\beta = -4.5$ when we use G2 EoS without hyperon. This makes deviation with GR.

4. Conclusion

From the previous results, deviation of GR which is in this case spontaneous scalarization appears in our new coupling function when using polytropic EoS at $\beta \approx -4.35$. Scalarization also appears when we use NL3 EoS either with hyperon or without hyperon. But, scalarization does not appear when we use G2 EoS with hyperon. The scalarization effect is larger when we use softer EoS. This result indicates the scalarization depends on the EoS of the NS.

ACKNOWLEDGMENT

We are partially supported by the UI's PITTA grant No. 2230/UN2.R3.1/HKP.05.00/2018 and No. 2267/UN2.R3.1/HKP.05,00/2018.

References

[1] Damour T and Esposito-Farese G 1996 Tensor - scalar gravity and binary pulsar experiments *Phys. Rev. D* **54** 1474
[2] Will C M 2014 The Confrontation between General Relativity and Experiment *Living Rev. Rel.* **17** 4
[3] Brans C and Dicke R H 1961 Mach's principle and a relativistic theory of gravitation *Phys. Rev.* **124** 925
[4] Novak J 1998 Neutron star transition to strong scalar field state in tensor scalar gravity *Phys. Rev. D* **58**
[5] Manolidis D 2014 Neutron Star Models in Alternative Theories of Gravity *All Theses and Dissertations Washington University*
[6] Faraoni V, Gunzig E and Nardone P 1998 Conformal transformations in classical gravitational theories and in cosmology *Fund. Cosmic Phys.* **20** 121
[7] Dabrowski M P, Garecki J and Blaschke D B 2008 Conformal transformations and conformal invariance in gravitation *Annalen Phys.* **18** 13
[8] Ramadhan H S 2011 Higher Dimensional Defect in Cosmology *Doctoral Dissertation Tufts University Massachusett USA*
[9] Prasetyo I 2017 Gravitational Field of Noncanonical Global Monopole: A Study of Their Black Holes and Compactifications *Master Thesis Universitas Indonesia*
[10] Shapiro S L and Teukolsky S A 2004 Black Holes, White Dwarfs, and Neutron Stars *Wiley-VCH*

Journal of Physics: Conference Series **1354** (2019) 012011 doi:10.1088/1742-6596/1354/1/012011

Testing Modified Gravity Theories by Constraining the Theories Parameter Using Radius and Mass Observations of Brown Dwarfs

A. S. Rosyadi, A. Sulaksono

Department of Physics, FMIPA, University of Indonesia, Kampus UI Depok, Indonesia

E-mail: adzilah.syahna@ui.ac.id, anto.sulaksono@sci.ui.ac.id

Abstract. Modified Gravity (MOG) theories modify Poisson equation in the Newtonian limit. This MOG is also important as test-bed to analyze whether General Relativity (GR) is consistent with observations. We investigate the mass and radius of Brown Dwarfs predicted by modified Poisson equation from Beyond Horndeski (BH) and Eddington-Inspired-Born-Infeld (EiBI) theories. Brown Dwarfs are sub-stellar objects that are difficult to be detected because of their low luminosity. One of the most important properties of brown dwarfs which recently observed by Johnson and Montet[1] are radii and masses. In this note, we derive the equation of state of the stars analytically, wherein electrons part, we use exact non-relativistic degenerate ideal Fermi gas result. We also derive modified hydrostatic stellar equation predicted by MOG theories and solve them numerically to obtain their profile and mass-radius relation for each parameter on both MOG theories. We compare our theoretical results with the ones obtained Johnson and Montet [1].

1. Introduction

MOG which succeed to overcome singularities problem in general relativity and served as alternatives to the existence of dark matter, need to be subjected to solar system or compact objects tests. In this research we focus on sub-stellar object such as brown dwarfs predictions of MOG (see Ref. [2] for similar study but focus on white dwarfs). Brown dwarfs are difficult to be detected because of their low luminosity. To study their luminosity, we need to know the properties of the objects, including radii and masses. From observation point of view, radii of brown dwarfs can be determined only if brown dwarfs is in transit period or existed in eclipse system. However, many progress in observation are reported recently (see Ref. [3] and the references therein). The first brown dwarf that is detected by Kepler data is found in LHS 6343 system. Johnson calculated mass of brown dwarf is $62.7 \pm 2.4 M_{Jup}$ and the corresponding radius is $0.833 \pm 0.021 R_{Jup}$. Montet also found brown dwarf with mass of $62.1 \pm 1.2 M_{Jup}$ and radii $0.782 \pm 0.013 R_{Jup}$. These results[1] depend on calculated mass with precision of 1.9% and depend on calculated radius with precision of 1.4% [1, 3]. We will compare these observation data to our theoritical results which are obtained use BH and EiBI theories in non-relativistic limit or each in a form of modified Poisson equation.

Content from this work may be used under the terms of the Creative Commons Attribution 3.0 licence. Any further distribution of this work must maintain attribution to the author(s) and the title of the work, journal citation and DOI.
Published under licence by IOP Publishing Ltd

2. Formalism

2.1. Modified Gravity Theories

The Vainshtein mechanism is only partially operational in BH family theories which has an incomplete supression of the fifth-force in the interior of extended sources due to the time-dependence of the cosmological field [4]. BH family theories is defined in action as

$$S = \int \sqrt{-g} M_{pl}^2 (\frac{R}{2} + X + \frac{\mathcal{L}_4}{\Lambda^4}) d^4x + S_M. \quad (1)$$

$M_{pl} = (8\pi G_N)^{-1}$ is planck mass, $X = -\frac{1}{2}\partial_a\phi\partial^b\phi$ and $\mathcal{L}_4 = X[(\Box\phi^2) - \phi_{ab}\phi^{ab}] - (\phi^a\phi^b\phi_{ab}\Box\phi - \phi^a\phi_{ab}\phi_c\phi^{cb})$. $F_5 = 2\alpha^2 F_N$ is a consistent solution of the equation and the force is unscreened. Through Vainshtein mechanism, few astrophysical signatures can be observed. One can see the detail derivation in Ref.[4], where the final modified Poisson equation of BH action is

$$\nabla^2\Phi(r) = 4\pi G_N \rho(r) + \frac{\gamma}{4} G_N \frac{d^3M(r)}{dr^3}. \quad (2)$$

γ is a dimensionless parameter which characterizes the deviation of of BH theory to general relativity (GR). Note that if $\gamma > 0$ in the new term in Eq. (2) then more mass is concentrated in stellar core [5].

We also investigate EiBI theory. The full action for EiBI theory[7],

$$S = \frac{1}{8\pi G_N \kappa} \int d^4x (\sqrt{|g_{ab} + \kappa R_{ab}(\Gamma)|} - \lambda\sqrt{-g}) + S_m(g_{ab}, \chi_m), \quad (3)$$

where $|.|$ represents the determinant. We will arrive at field equation after varying EiBI action,

$$\sqrt{-q} q^{ab} = \lambda\sqrt{-g} g^{ab} - 8\pi G_N \kappa \sqrt{-g} T^{(m)ab}, \quad (4)$$

where q^{ab} is the inverse of q_{ab} and $T^{(m)}_{ab}$ is energy-momentum tensor. The lowest correction Einstein equation in Newtonian limit will give modified Poisson equation. The modified Poisson equation for EiBI is expressed by [6],

$$\nabla^2\Phi(r) = 4\pi G_N \rho(r) + \frac{\kappa}{4} \nabla^2 \rho(r), \quad (5)$$

where the correction term can work as repulsive or attractive force depending on the sign of the parameter κ and it can be shown that this parrameter corresponds to effective polytropic equation of state parameter K through $K = \frac{\kappa}{8}$ relation [6].

2.2. Equation of State

Brown dwarf fail to have a stable hydrogen burning sequence and instead derive their stability from electron degeneracy pressure [8]. We integrate the Fermi-Dirac integral exactly using the polylogarithm functions $\mathrm{Li}_s(x)$. The electrons are mainly non-relativistic due to relatively low temperature and density. The expression for the pressure of a degenerate Fermi gas at finite temperature is

$$\begin{aligned} P_F &= \frac{2}{5}\alpha\mu^{5/2} - \frac{1}{8}\alpha\mu^{3/2} Log(1 + e^{-\beta\mu}) + \frac{3}{2}\alpha\beta^{-2}\mu^{1/2}\frac{\pi^2}{6} \\ &+ \frac{3}{4}\alpha\beta^{-2}\mu^{1/2} Li_2(-e^{-\beta\mu}) - \frac{3}{4}\alpha\beta^{-3}\mu^{-1/2} Li_3(-e^{-\beta\mu}) \\ &+ \left(\frac{\pi^2}{6} - \frac{9}{4}\right)\alpha\beta^{-1}\mu^{3/2} + \left(\frac{9}{8} - \frac{\pi^2}{3}\right)\alpha\beta^{-2}\mu^{1/2} + \left(\frac{\pi^2}{2}\frac{3}{8}\right)\alpha\beta^{-3}\mu^{-1/2} \\ &+ \frac{1}{8}\alpha\mu^{5/2}. \end{aligned} \quad (6)$$

where $\alpha = (2/3)(4\pi(2m)^{3/2}/(2\pi\hbar)^3)$, $\beta = (k_B T)^{-1}$. We introduce degeneracy parameter ψ as central temperature of brown dwarf. It has the same order of magnitude as the electron Fermi Temperature,

$$\psi = \frac{k_B T}{\mu_F} = \frac{2m_e}{\hbar^2}\left(\frac{m_H \mu_e}{3\pi^2 \rho}\right)^{2/3} k_B T, \tag{7}$$

where μ_F is the electron Fermi energy in the degenerate limit. $\mu_e = 1.143$ is a fixed parameter of number of baryons per electron [5].

The interior of brown dwarf is composed by electrons and ions. Thus, the total pressure comes from both contributions i.e., $P = P_F + P_{ion}$. The pressure due to ions for an ionized hydrogen gas can be expressed by $P_{ion} = \frac{k\rho T}{\mu_1 m_H}$. The total equation of state for brown dwarf pressure is then

$$\begin{aligned} P &= \frac{2}{5}\alpha A^{5/2}\left(\frac{\rho}{\mu_e}\right)^{5/3}\left\{1 - \frac{5}{16}\psi Log(1+e^{-\frac{1}{\psi}}) + \frac{15}{8}\psi^2\left(\frac{\pi^2}{3} + Li_2(-e^{-\frac{1}{\psi}})\right)\right. \\ &- \frac{15}{16}\psi^3 Li_3(-e^{-\frac{1}{\psi}}) + \frac{5}{2}\left(\frac{\pi^2}{6} - \frac{9}{4}\psi\right) \\ &+ \left.\frac{5}{2}\left(\frac{9}{8} - \frac{\pi^2}{3}\right)\psi^2 + \frac{5}{2}\left(\frac{3\pi^2}{16}\right)\psi^3 + \frac{5}{16} + \alpha\psi\right\}, \end{aligned} \tag{8}$$

where $\alpha = \frac{5\mu_e}{2\mu_1}$.

We can express the relation in Eq. (8) in a polytropic equation of state relation as $P = K\rho^{5/3}$ which 5/3 is related to polytropic index $n = 3/2$, if we define

$$K = \frac{(3\pi^2)^{2/3}\hbar^2}{5 m_e m_H^{5/3} \mu_e^{5/3}}(1 + \gamma + \alpha\psi), \tag{9}$$

where $\alpha = (2/3)(4\pi(2m)^{3/2}/(2\pi\hbar)^3)$,$\psi$ is a degeneracy parameter and correction term γ is expressed by,

$$\begin{aligned} \gamma &= -\frac{5}{16}\psi Log(1+e^{-1/\psi}) + \frac{15}{8}\psi^2\left(\frac{\pi^2}{3} + Li_2(-e^{-1/\psi})\right) \\ &- \frac{15}{16}\psi^3 Li_3(-e^{-1/\psi}) + \frac{5}{2}\left(\frac{\pi^2}{6} - \frac{9}{4}\right)\psi \\ &+ \frac{5}{2}\left(\frac{9}{8} - \frac{\pi^2}{3}\right)\psi^2 + \frac{5}{2}\left(\frac{\pi^2}{2}\frac{3}{8}\right)\psi^3 + \frac{5}{16}. \end{aligned} \tag{10}$$

This constant depends on central density of stellar interior which rules the distribution of density and temperature in all points of brown dwarf.

2.3. Newtonian Limit of Hydrostatic Equilibrium Stellar Equation

Eq. (11) of stellar structure can be derived by considering the cylindrical slab in the star with two forces, one is from inside to outside which is known as electron degeneracy and the other is the gravitational force from the outside to inside. Both is in equilibrium. While Eq. (12) is derived from mass continuity. The explicit form of both equations are

$$\frac{dP(r)}{dr} = -\frac{d\phi(r)}{dr}\rho(r) \tag{11}$$

$$\frac{dM(r)}{dr} = 4\pi r^2 \rho(r) \tag{12}$$

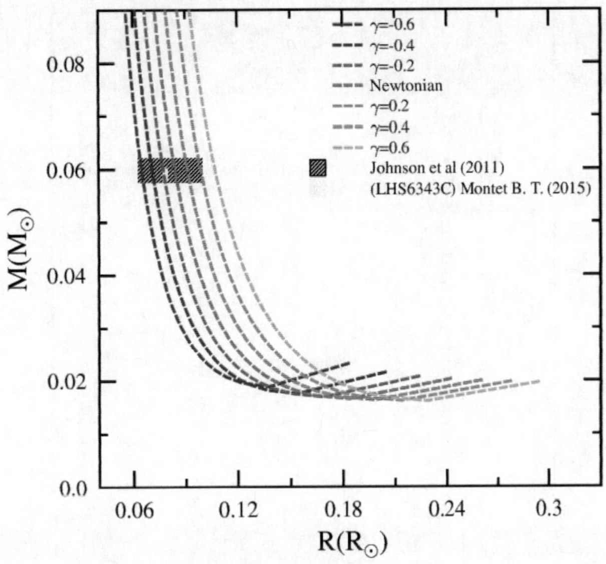

Figure 1. Beyond Horndeski Gravity on Mass and Radius Relation

We can use this equation to solve the stellar equation of Brown Dwarfs. We can obtain the modified Poisson equation of two modified gravity theories from subsection (2.3) from Eqs. (11) and (12). The modified Poisson equation for BH theory is

$$\frac{d\rho(r)}{dr} = -\frac{1}{K\Gamma\rho(r)^{\Gamma-1} + \gamma G_N \pi r^2 \rho(r)} \left\{ \frac{G_N M(r)}{r^2}\rho(r) + 2\pi\gamma G_N r \rho(r)^2 \right\}, \tag{13}$$

where γ is a dimensionless parameter of BH theory. We also can obtain the stellar equation for EiBI theory as

$$\frac{d\rho(r)}{dr} = -\frac{1}{K\Gamma\rho(r)^{\Gamma-1} + \frac{\kappa}{4}\rho(r)} \left\{ \frac{G_N M(r)}{r^2}\rho(r) \right\}, \tag{14}$$

where κ is a dimensionless parameter of EiBI theory.

3. Result and Discussion

We solve differential equation from subsection (2.3) numerically using fouth order Runga-Kutta (RK4) dimensionless parameter with initial boundary condition $M(0) = 0$ and $\rho(0) = \rho_c$ where ρ_c is central density of brown dwarf ($10^4 - 10^6 kg/m^3$) [9] to obtain mass-radius relation for brown dwarfs.

For BH Gravity theory, γ parameter can be positive or negative. For $\gamma > 0$, the correction term in hydrostatic equation works as repulsive force, while for For $\gamma < 0$, it works as attractive force. As the result in modified mass and radius relation, the radius is increased by increasing γ value.

This situation is quite similar to those of EiBI gravity theory. The similarity can be ssen obviously by comparing Fig. (1) of BH and Fig. (2) of EiBI. Parameter κ in EiBI can also work as attractive or repulsive force depending on κ sign.

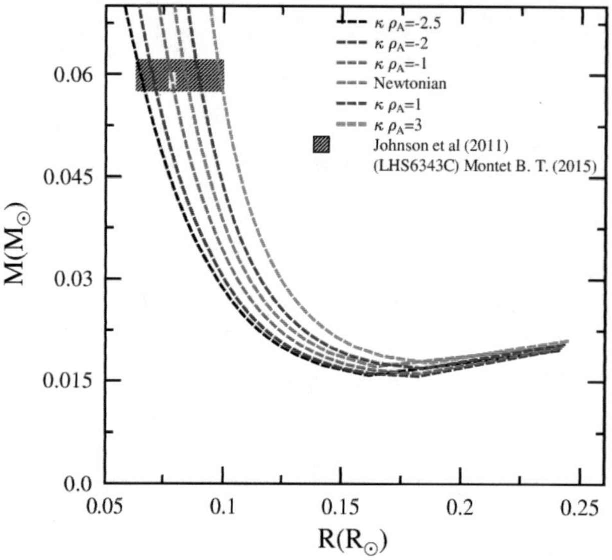

Figure 2. Eddington Inspired Born-Infeld Gravity on Mass and Radius Relation

Interestingly, Both theories predict the minimum mass of brown dwarf. Note that the minimum mass of Hydrogen Burning (MMHB) has been studied quite detail in Ref.[5]. Furthermore, we obtain the allowed parameter range between $-0.6 \leq \gamma \leq 0.4$ using observational constraints of Johnson *et. al* and thighter constraints for $\gamma = -0.2$ using Montet *et. al* data (LHS6343C). Note that in BH theory, Sakstein obtain the constraint limit for $\gamma < -2/3$ in order brown dwarfs have a stabilized stellar configuration [5]. In the case of EiBI theory, We can obtain by using constraint from Johnson *et. al* the parameter range between $-2.5 \leq \kappa\rho_A \leq 3$. Here $\rho_A = \rho_\odot \times 10^9$. We also obtain tighter constraint for $\kappa\rho_A = -1$ by using Montet *et al* data (LHS6343C).

4. Conclusion

The free parameter of two modified gravity theories gives significant effects to masses and radii relation of brown dwarfs. These are the consequences of correction term in modified Poisson equations that can work as repulsive or attractive force term so that it can strength or weaken gravitational field in stellar objects in general. We can obtain the limit of free parameter of both theories from the brown dwarf observational masses and radii. The corresponding masses and radii constraints by Johnson *et. al* and Montet *et al* which are shown in Figs. (1) and (2) can tightly constraint The free parameter of both modified gravity theories.

ACKNOWLEDGMENT

We are partially supported by the UI's PITTA grant No. 2230/UN2.R3.1/HKP.05.00/2018

Journal of Physics: Conference Series **1354** (2019) 012011 doi:10.1088/1742-6596/1354/1/012011

References

[1] Montet et al and Johnson et al, "Characterizing The Cool KOIs. VII. Refined Physical Properties of The Transiting Brown Dwarf LHS 6343 C," The Astrophysical Journal, 800:134(11pp)(2015).

[2] S. Banarjee, S. Shankar, and T. P. Singh, "Constraints on Modified Gravity Models from White Dwarfs," Journal of Cosmology and Astroparticle Physics (JCAP) (2017).

[3] D. Bayliss, "EPIC 20170247b: A Transiting Brown Dwarf from K2 IN a 41 Day Orbit," The Astronomical Journal 153.1 (2016).

[4] K. Koyama and J. Sakstein, "Astrophysical Probes of The Vainstein Mechanism: Stars and Galaxies," Phys. Rev. D. D91,124066 (2015).

[5] J. Sakstein, "Hydrogen Burning in Low Mass Stars Constrains Scalar-Tensor Theories of Gravity," Phys. Rev. Lett. 115,201101 (2015).

[6] P. Pani, T. Delsate, and V. Cardoso, "Eddington-inspired Born Infeld Gravity: Phenomenology of Nonlinear Gravity Matter Coupling," Phys. Rev. D. 85,084020 (2012).

[7] M. Banados and P. G. Ferrerira, "Eddington's theory of gravity and its progeny," Phys. Rev. Lett. 105,011101 [arxiv: 1006.1769v2] (2010).

[8] S. Auddy, S. Basu, and S.R. Valluri, "Analytical Models of Brown Dwarfs and The Substellar Mass Limit," [arXiv:1607.0433v1](2016).

[9] A. Burrows and J. Liebert, "The Science of Brown Dwarfs," Reviews of Modern Physics, vol. 65, no.2, p. L199 (1998).

Journal of Physics: Conference Series **1354** (2019) 012012 doi:10.1088/1742-6596/1354/1/012012

Tidal Love Numbers of Fermionic Dark Stars

A. B. Wahidin and A. Sulaksono

Departemen Fisika, Fakultas Matematika dan Ilmu Pengetahuan Alam, Universitas Indonesia, Depok, Indonesia

E-mail: alka.budi@sci.ui.ac.id, anto.sulaksono@sci.ui.ac.id

Abstract. Dark star is a hypothetical stellar compact object composed of dark particles which clumped together due to self-interaction among dark particles. Tidal Love numbers from binary stars are potential observables for studying the internal structure of the stars. The Tidal Love numbers can be deduced from the gravitational wave of the coalescence or inspiral of binary stars. The ground base interferometer can measure the corresponding gravitational wave. In this work, we study the equation of state, mass-radius relation, electric, magnetic and surficial tidal love numbers of dark stars. We obtain that Tidal Love numbers are not too sensitive to study the role of scalar boson exchange and the impact of the mass of the dark particle in dark matter equation of state.

1. Introduction

It is recently reported that cold collisionless dark matter (CCDM) as the main candidate of dark matter is not too compatible with some astrophysical observations. There are three main problems of CCDM paradigm [1, 2]. The flat density core in dwarf galaxies observations is contradicted with numerical simulation result from CCDM. CCDM also predicts a larger number of satellite galaxies in the Milky Way that have been observed so far. CCDM also predicts massive dwarf galaxies that are too big not to be observed.

Recent discussions have shown that dark matter (DM) with self-interaction can explain not only diverse galactic rotational curves [3] but also can resolve CCDM problems in appropriate range of DM self-interaction strength i.e., in the range of the ratio of total cross section σ to dark particle mass m between [1, 2] 0.1 cm^2/g $< \sigma/m <$ 10 cm^2/g. The consequence of the existence of DM composed by self-interaction dark particles is the corresponding dark particles may clump together to form a compact object called a dark star. It is also reported that dark stars are not compact enough to mimic black holes [2]. Therefore, dark stars can be distinguishable from black holes when we observe the dark sector of the universe.

The new era of astronomical observations has begun since observations of the binary black hole coalescence GW170814 [4] and the binary neutron star inspiral GW170817 [5] by LIGO and Virgo collaborations. Tidal effect of binary stars is measurable during the early part of evolution before coalescence because the gravitation waveform in this stage is relatively clean [6, 7]. Several studies have shown that the measurement of tidal deformation can be used to extract the information about internal structure of neutron stars [8, 9, 10, 11, 12, 13, 14].

If dark stars exist, the study of the tidal deformation of binary dark stars may shade a light to provide more understanding of the nature of DM. In this paper we study electric tidal love

Content from this work may be used under the terms of the Creative Commons Attribution 3.0 licence. Any further distribution of this work must maintain attribution to the author(s) and the title of the work, journal citation and DOI.
Published under licence by IOP Publishing Ltd

number k_l and surficial tidal love number h_l for $l = 2, 3, 4$, while for magnetic tidal love number j_2 only for $l = 2$ predicted by fermionic dark stars.

2. Formalism

To describe fermionic DM, we use Lagrangian self-interacting fermionic dark particle where ψ, ϕ, V^μ denotes to dark particle, scalar boson and vector boson fields, respectively and m, m_s, m_v denote their masses with coupling strength g_s for scalar boson and g_v for vector boson [15, 17, 18].

$$\mathcal{L} = \bar{\psi}\left[\gamma_\mu(i\partial^\mu - g_v V^\mu) - (m - g_s\phi)\right]\psi + \frac{1}{2}(\partial_\mu\phi\partial^\mu\phi - m_s^2\phi^2) - \frac{1}{4}V^{\mu\nu}V_{\mu\nu} + \frac{1}{2}m_v^2 V^\mu V_\mu \quad (1)$$

where $V^{\mu\nu} = \partial^\mu V^\nu - \partial^\nu V^\mu$. Using standard relativistic mean field (RMF) approximation, we can obtain equation of motion as

$$\left[(i\gamma_\mu\partial^\mu - m) + g_s\phi_0 + g_v V_0\gamma_0\right]\psi = 0, \quad (2)$$

$$m_s^2\phi_0 = g_s\langle\bar{\psi}\psi\rangle, \quad (3)$$

$$m_v^2 V_0 = g_v\langle\bar{\psi}\gamma^0\psi\rangle, \quad (4)$$

and for energy-momentum tensor is given by

$$T^{\mu\nu}_{RMF} = \bar{\psi}i\gamma^\mu\partial^\nu\psi - \eta^{\mu\nu}\left[-\frac{1}{2}m_s^2\phi_0^2 + \frac{1}{2}m_v^2V_0^2\right]. \quad (5)$$

From the later equation, we can further obtain energy density as

$$\epsilon = \langle T^{00}\rangle = \langle\bar{\psi}i\gamma^0\partial^0\psi\rangle + \frac{1}{2}m_s^2\phi_0^2 - \frac{1}{2}m_v^2V_0^2, \quad (6)$$

and pressure as

$$P = \frac{1}{3}\langle T^{ii}\rangle = \frac{1}{3}\langle\bar{\psi}i\gamma^i\partial^i\psi\rangle - \frac{1}{2}m_s^2\phi_0^2 + \frac{1}{2}m_v^2V_0^2. \quad (7)$$

The corresponding expectation values can be also easily calculated and the results are [15, 18]

$$\langle\bar{\psi}i\gamma^0\partial^0\psi\rangle = g_vV_0\rho + \frac{\gamma}{(2\pi)^3}\int_0^{k_F} d\mathbf{k}\sqrt{\mathbf{k}^2 + m^{*2}}, \quad (8)$$

$$\langle\bar{\psi}i\gamma^i\partial^i\psi\rangle = \frac{\gamma}{(2\pi)^3}\int_0^{k_F} d\mathbf{k}\frac{\mathbf{k}^2}{\sqrt{\mathbf{k}^2 + m^{*2}}}, \quad (9)$$

$$\langle\bar{\psi}\psi\rangle = \frac{\gamma}{2\pi^2}\int_0^{k_F} dk\frac{k^2m^*}{\sqrt{k^2 + m^{*2}}}, \quad (10)$$

and

$$\langle\bar{\psi}\gamma^0\psi\rangle = \left\langle\psi^\dagger\psi\right\rangle = \sum_s\frac{1}{(2\pi)^3}\int_0^{k_F} d^3k = \frac{\gamma}{6\pi^2}k_F^3 \equiv \rho, \quad (11)$$

where $m^* = m - g_s\phi$. Therefore, the boson fields can be expressed as

$$\phi_0 = \frac{g_s}{m_s^2}\langle\bar{\psi}\psi\rangle = \frac{g_s}{m_s^2}\frac{\gamma}{2\pi^2}\int_0^{k_F} dk\frac{k^2m^*}{\sqrt{k^2 + m^{*2}}}, \quad (12)$$

$$V_0 = \frac{g_v}{m_v^2}\left\langle\psi^\dagger\psi\right\rangle = \frac{g_v}{m_v^2}\rho, \quad (13)$$

Table 1. Parameters used in this work to generate the DM EOS. $m_v = m_s \equiv$ 10 MeV

EOS	m	α_v	α_s
I	1 GeV	3.0×10^{-3}	2.0×10^{-3}
II		0.7×10^{-3}	0.3×10^{-3}
III	2 GeV	3.0×10^{-3}	2.0×10^{-3}
IV		0.7×10^{-3}	0.3×10^{-3}

$$m^* = m - \frac{g_s^2}{m_s^2}\int_0^{k_F} dk \frac{k^2 m^*}{\sqrt{k^2+m^{*2}}}, \tag{14}$$

$$\epsilon = \frac{1}{2}m_s^2\phi_0^2 + \frac{1}{2}m_v^2V_0^2 + \frac{\gamma}{2\pi^2}\int_0^{k_F} dk k^2\sqrt{k^2+m^{*2}}, \tag{15}$$

$$P = -\frac{1}{2}m_s^2\phi_0^2 + \frac{1}{2}m_v^2V_0^2 + \frac{1}{3}\left(\frac{\gamma}{2\pi^2}\int_0^{k_F} dk \frac{k^4}{\sqrt{k^2+m^{*2}}}\right). \tag{16}$$

After solving those integrals, Eqs. (12-16) can be written in more compact forms. The corresponding final expressions for effective mass, energy density and pressure are

$$m^* = m - \frac{g_s^2}{m_s^2}\frac{m^{*3}x^3}{2\pi^2}\chi(x), \tag{17}$$

$$\epsilon = m^{*4}\xi(x) + \left[\frac{2}{9\pi^3}m^{*6}x^6\right]\left[\frac{\alpha_v}{m_v^2} + \frac{9}{4}\frac{\alpha_s}{m_s^2}(\chi(x))^2\right], \tag{18}$$

$$P = m^{*4}\Psi(x) + \left[\frac{2}{9\pi^3}m^{*6}x^6\right]\left[\frac{\alpha_v}{m_v^2} - \frac{9}{4}\frac{\alpha_s}{m_s^2}(\chi(x))^2\right]. \tag{19}$$

where $\alpha_i \equiv g_i^2/4\pi$, and exact form of those integrals are

$$\begin{aligned}
\chi(x) &\equiv \frac{1}{x^3}\left[x\sqrt{x^2+1} - \ln\left(x+\sqrt{x^2+1}\right)\right].\\
\xi(x) &\equiv \frac{1}{8\pi^2}\left[x\sqrt{x^2+1}\,(2x^2+1) - \ln(x+\sqrt{x^2+1})\right],\\
\Psi(x) &\equiv \frac{1}{8\pi^2}\left[x\sqrt{x^2+1}\,(\tfrac{2}{3}x^2-1) + \ln(x+\sqrt{x^2+1})\right],
\end{aligned} \tag{20}$$

Note that $\gamma = 2$ for spin degeneracy factor and $x = k_F/m^*$. Eqs. (17-20) are used to generate equation of state (EOS) as parametric function of x. In calculation, we use parameters shown in table 1. Those parameters yield ratios of cross section to dark particle mass which are still inside the 0.1 cm^2/g$< \sigma/m <$10 cm^2/g constraint. The EOS results are shown in figure 1.

Journal of Physics: Conference Series **1354** (2019) 012012 doi:10.1088/1742-6596/1354/1/012012

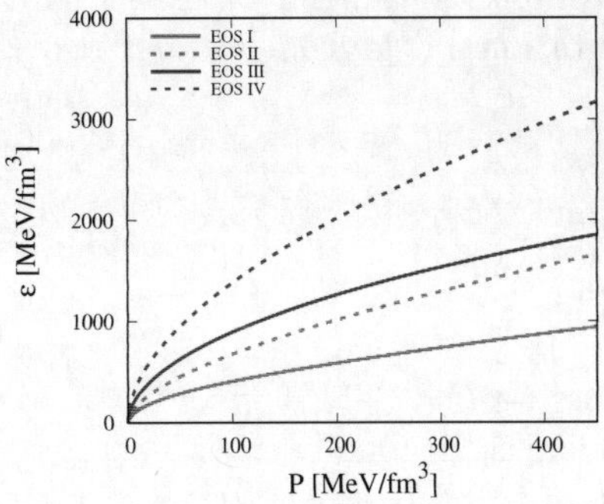

Figure 1. EOSs for each parameter set used in this work.

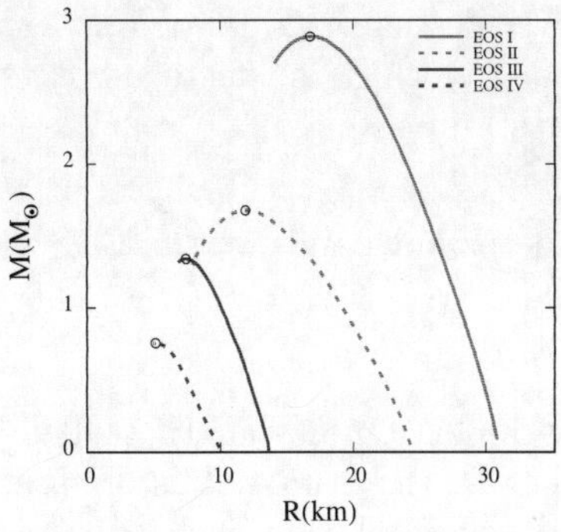

Figure 2. Mass-Radius relation for each EOS used in this work. $M_\odot$ is solar mass.

3. Mass and Radius

Mass and radius are macroscopic observable that characterize astrophysical compact object. We can calculate them by solving Tolman-Oppenhaimer-Volkoff(TOV) equations bellow

$$\frac{dP(r)}{dr} = -\frac{GM(r)\epsilon(r)}{r^2}\left(1+\frac{P(r)}{\epsilon(r)}\right)\left(1+\frac{4\pi r^3 P(r)}{M(r)}\right)\left(1-\frac{2GM(r)}{r}\right)^{-1}, \quad (21)$$

$$\frac{dM(r)}{dr} = 4\pi\epsilon(r)r^2, \quad (22)$$

where $M(r)$ denotes the profile of a stellar mass, r denotes radius, G is gravitational constant. These equations need EOS as an input. In this work, we plot $\epsilon(r)$ and $P(r)$ from Eqs. (18) and (19) as parametric data so we get the plot line in figure 1.To make our numerical problem easier, we can aprroximate exact function of $\epsilon(P)$ with some polinomial using fitting method. This method is more efficient rather than call $\epsilon(x)$ and $P(x)$ for each iteration which is contains m^* as a self consistent function that need a lot of process to get the result. So TOV equations become

$$\frac{dP(r)}{dr} = -\frac{GM(r)\epsilon(P(r))}{r^2}\left(1+\frac{P(r)}{\epsilon(P(r))}\right)\left(1+\frac{4\pi r^3 P(r)}{M(r)}\right)\left(1-\frac{2GM(r)}{r}\right)^{-1}, \quad (23)$$

$$\frac{dM(r)}{dr} = 4\pi\epsilon(P(r))r^2, \quad (24)$$

For solving TOV equations numerically, Runge-Kutta method is used which is require initial values in the center of the stars. We use $P(0) = P_c$ and $M(0) \approx 0$ as pressure and mass at the center of the star. We don't know the exact value of P_c so we variate that. We determine the star radius R when $P(R) \approx 0$ (Pressure in the surface of the stars approach zero), and in this point we obtain the star total mass $M = M(R)$ and radius R of the stars. As we mention earlier, we variate P_c until maximum mass is reached. After reaching maximum mass, larger P_c value tends to lower the star masses. In this region, the star is unstable .The mass-radius results are shown in figure 2.

4. Tidal Love Numbers

There is a possibility to extract information of the internal structure of astronomical compact stars via gravitational wave through tidal deformation properties of the stars in inspiral binary stars during coalescence. Tidal deformability is a response parameter of mass quadrupole moment of a star Q_{ij} which is affected by external gravitational field from another star (tidal field) $\mathcal{E}_{ij}$ of binary stars.

$$Q_{ij} = -\lambda\mathcal{E}_{ij}, \tag{25}$$

where parameter λ depends on the EOS via a relation

$$\lambda = \frac{2}{3}k_2 R^5 \tag{26}$$

where k_2 are dimensionless quantity called love number for $l = 2$. This effect becomes important over time because the orbit radius of binary stars decrease gradually due to energy loss due to gravitational waves. When each star becomes closer to each other, gravitational attraction becomes stronger and stars will be deformed. So quantity k_2 so on for higher order k_3 and k_4 provide important information which is obtained from gravitational wave[16].

To calculate love numbers k_l for $(l = 2, 3, 4)$, we solve differential equation below [16]

$$r\frac{dy_l(r)}{dr} + y_l(r)^2 + y_l(r)F(r) + r^2 Q_l(r) = 0, \tag{27}$$

here $l = 2, 3, 4....$ Note that the explicit form of $F(r)$ and $Q_l(r)$ is

$$F(r) = 1 - 4\pi r^2 G(\epsilon(r) - P(r))\left(1 - 2\frac{GM(r)}{r}\right)^{-1}, \tag{28}$$

$$\begin{aligned} Q_l(r) &= \left[4\pi r^2 G\left(5\epsilon(r) + 9P(r) + \frac{\epsilon(r) + P(r)}{(\partial P/\partial\epsilon)(r)}\right) - l(l+1)\right]\left(1 - 2\frac{GM(r)}{r}\right)^{-1} \\ &- 4\left(\frac{GM(r)}{r}\right)^2\left(1 + \frac{4\pi r^3 P}{M(r)}\right)^2\left(1 - 2\frac{GM(r)}{r}\right)^{-2}. \end{aligned} \tag{29}$$

Because those equation need variable with dependence of r, we should solve equation (27) and TOV equation simultaneously with Runge Kutta method. Initial value $y_l(0) = l$ where $l = 2, 3, 4$ is used.

When we know the value of $y_l = y_l(R)$, we can calculate k_l from following equations

$$\begin{aligned} k_2 = \frac{8}{5}(1-2C)^2 C^5[2C(y_2-1) - y_2 + 2]\Big\{&2C(4(y_2+1)C^4 + (6y_2-4)C^3 + (26-22y_2)C^2 \\ &+3(5y_2-8)C - 3y_2 + 6) - 3(1-2C)^2(2C(y_2-1) - y_2 + 2)\ln\left(\frac{1}{1-2C}\right)\Big\}^{-1} \end{aligned} \tag{30}$$

$$\begin{aligned} k_3 = \frac{8}{7}(1-2C)^2 C^7[2(y_3-1)C^2 - 3(y_3-2)C + y_3 - 3]\Big\{&2C[4(y_3+1)C^5 + 2(9y_3-2)C^4 \\ -20(7y_3-9)C^3 + 5(37y_3-72)C^2 - 45(2y_3-5)C + 15(y_3-3)] &- 15(1-2C)^2 \\ \times(2(y_3-1)C^2 - 3(y_3-2)C + y_3 - 3)\ln&\left(\frac{1}{1-2C}\right)\Big\}^{-1} \end{aligned} \tag{31}$$

$$\begin{aligned} k_4 = \frac{32}{147}&(1-2C)^2 C^9[12(y_4-1)C^3 - 34(y_4-2)C^2 + 28(y_4-3)C - 7(y_4-4)] \\ &\times\Big\{2C[8(y_4+1)C^6 + (68y_4-8)C^5 + (1284-996y_4)C^4 + 40(55y_4-116)C^3 \\ &+(5360-1910y_4)C^2 + 105(7y_4-24)C - 105(y_4-4)] - 15(1-2C)^2 \\ &\times[12(y_4-1)C^3 - 34(y_4-2)C^2 + 28(y_4-3)C - 7(y_4-4)]\ln\left(\frac{1}{1-2C}\right)\Big\}^{-1}, \end{aligned} \tag{32}$$

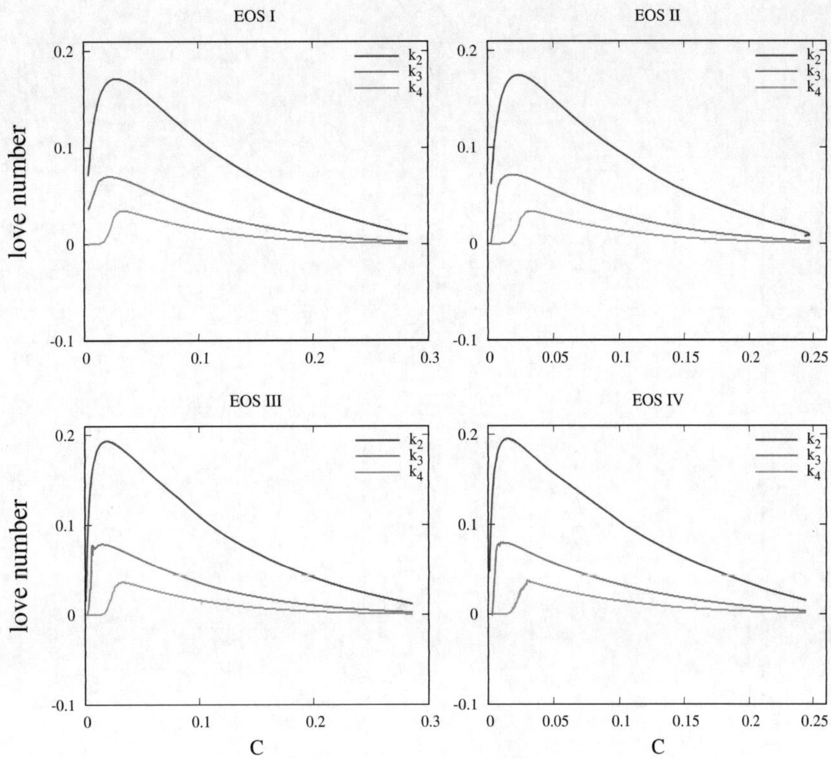

Figure 3. Electric Love numbers for each EOS used.

where $C = GM/R$ is a compactness. We calculate the surfical love number h_l which describe deformation of the body's surface in a multipole expansion. For a perfect fluid, the relation between surfical love number h_l and tidal love number k_l is given by [16]

$$h_l = \Gamma_1 + 2\Gamma_2 k_l, \tag{33}$$

where

$$\Gamma_1 = \frac{l+1}{l-1}(1-C)F(-l,-l,-2l;2C) - \frac{2}{l-1}F(-l,-l-1,-2l;2C) \tag{34}$$

$$\Gamma_2 = \frac{l}{l+2}(1-C)F(l+1,l+1,2l+2;2C) + \frac{2}{l+2}F(l+1,l,2l+2;2C). \tag{35}$$

Here $F(a,b,c;z)$ is the hyper-geometric function. We also calculate the magnetic tidal love number j_l, but only for $l = 2$ case with the final expression is given as[16]

$$j_2 = \left\{96C^5(2C-1)(y_2-3)\right\}\Big\{5(2C(12(y_2+1)C^4 + 2(y_2-3)C^3 + 2(y_2-3)C^2 + 3(y_2-3)C - 3y_2 + 9) + 3(2C-1)(y_2-3)\ln(1-2C))\Big\}^{-1}. \tag{36}$$

5. Result and Discussion

Solving Eqs.(17-20) by using the parameter sets shown in table 1, we can obtain the EOSs which are shown in figure 1. It can be observed that EOS IV and EOS II are stiffer than those of EOS

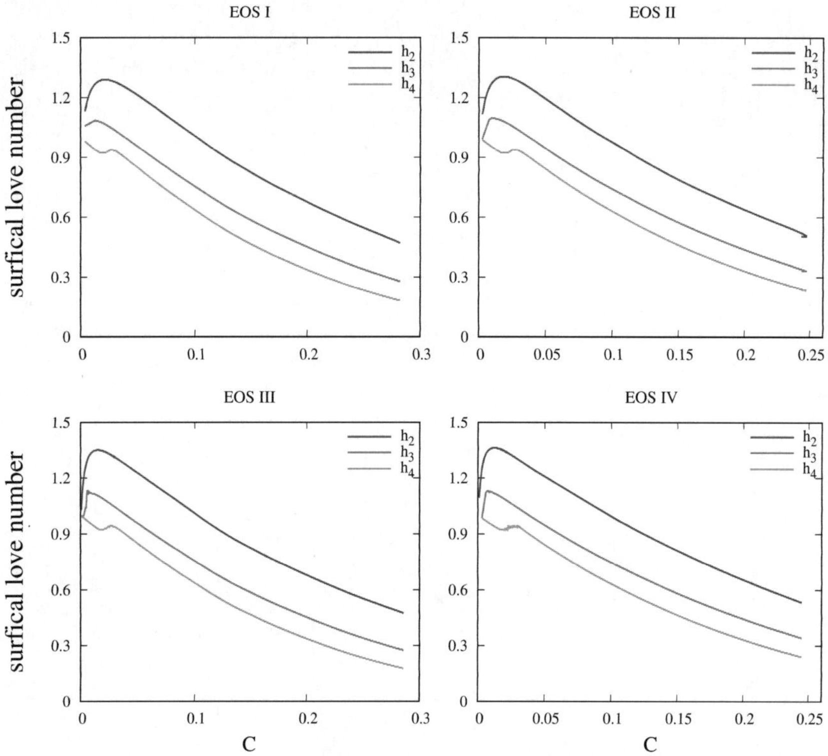

Figure 4. Surfical Love numbers for each EOS used.

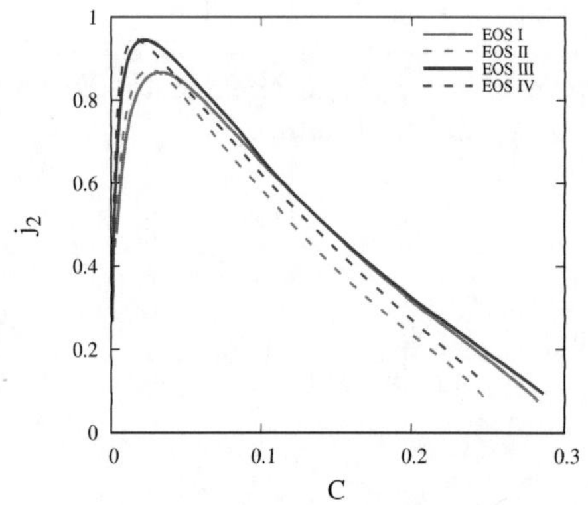

Figure 5. Magnetic tidal Love number for each EOS used.

III and EOS I at high densities but they have the same behavior at low densities. It can be also seen that for heavier DM mass, the EOS tends to be softer than that of small DM mass. The mass-radius relation results of the corresponding EOSs are shown in figure 2. It is obvious that the maximum mass reduces significantly if the DM mass increases and the EOS is softened.

We solve Eqs.(27-29) alongside with TOV equations to obtain the electric love number k_l

for ($l = 2, 3, 4$). $l = 2, 3, 4$ represent quadruple, octupole and hexadecapole deformations, respectively. In figure3, we show the electric tidal Love numbers predicted by the EOSs in table 1. For heavier dark particle mass k_2 peak tends to shift slightly to the larger compactness and the k_2 peak magnitude becomes higher. This result can be understood because if the EOS is softer, the star can deform more easily. Note that the stiffness of the EOS can be tuned by increasing or decreasing dark particle mass or by adjusting the boson coupling constants. We also can see that for larger compactness, all k_l tend to go zero as compactness is closer the compactness of black hole $C = 0.5$. We can also see this similar pattern for higher order Love numbers.

Using Eq. (33) we can also calculate the surficial Love numbers where the results are shown in figure 4. It can be observed that similar to the ones of electric tidal Love numbers, the difference in EOS is not also significantly reflected in surficial Love numbers.

It can be seen in figure 5, for magnetic tidal love numbers, the peak shift depends strongly to the dark particle mass. However, at relative high compactness region, the magnetic tidal Love numbers with the same coupling constant strength tend to coincide one another.

6. Conclusion

In this work, we study the effect of dark particle mass and scalar boson coupling exchange to the EOS, mass-radius relation, and tidal Love numbers properties of fermionic dark stars. We calculate and discuss the electric tidal Love numbers, surficial tidal Love numbers, and magnetic tidal Love numbers. We obtain that tidal love numbers are not too sensitive in respect to the differences of the used EOSs. Therefore, it seems that they are not quite sensitive to investigate the dark particle mass and scalar boson coupling of fermionic dark stars.

7. Acknowledgments

We are partially supported by the UIs PITTA grant No. 2231/UN2.R3.1/HKP.05.00/2018.

References

[1] Kouvaris C and Nielsen N G 2015 *Phys. Rev.* D **92** 063526
[2] Maselli A, Pnigouras P, Nielsen N G, Kouvaris C and Kokkotas K D 2017 *Phys. Rev.* D **96** 023005
[3] Kamada A, Kaplinghat M, Pace A B and Yu H B 2017 *Phys. Rev.* Lett **119** 111102
[4] Abbott B P *et al.* 2017 *Phys. Rev.* Lett **119** 141101
[5] Abbott B P *et al.* 2017 *Phys. Rev.* Lett **119** 161101
[6] Flanagan E E and Hinderer T 2008 *Phys. Rev.* D **77** 021502
[7] Hinderer T 2008 *Astrophys. J.* **677** 1216
[8] Baiotti L, Damour T, Giacomazzo B, Nagar A and Rezzolla L 2011 *Phys. Rev.* D **84** 024017
[9] Vines J, Flanagan E E and Hinderer T 2011 *Phys. Rev.* D **83** 084051
[10] Pannarale F, Rezzolla L, Ohme F and Read J S 2011 *Phys. Rev.* D **84** 104017
[11] Lackey B D, Kyutoku K, Shibata M, Brady P R and Friedman J L 2012 *Phys. Rev.* D **85** 044061
[12] Damour T, Nagar A and Villain L 2012 *Phys. Rev.* **85** 123007
[13] Hinderer T *et al.* 2016 *Phys. Rev.* Lett **116** 181101
[14] Essick R, Vitale S and Weinberg N N 2016 *Phys. Rev.* D **94** 103012
[15] Diener J W P 2008 *Relativistic mean-field theory applied to the study of neutron star properties* (Stellenbosch University)
[16] Kumar B, Biswal S K and Patra S K 2017 *Phys. Rev.* C **95** 015801
[17] Xiang Q F, Jiang W Z, Zhang D R and Yang R Y *Phys. Rev.* C **89** 025803 *Phys. Rev. C* **89**, 025803 (2014) .
[18] Glendenning N K 1997 *Compact stars: Nuclear physics, particle physics, and general relativity* (New York, USA: Springer)

Journal of Physics: Conference Series **1354** (2019) 012013 doi:10.1088/1742-6596/1354/1/012013

Multiwavelength analysis of nearby ultraluminous x-ray sources (ULXs) and their environment

K Vierdayanti, S A Khairunnisa and H Wulandari

Astronomy Research Division, Faculty of Mathematics and Natural Sciences, Institut Teknologi Bandung, Jl. Ganesha 10, Bandung 40132, Indonesia

E-mail: kiki@as.itb.ac.id

Abstract. Ultraluminous X-ray sources (ULXs) are known as extragalactic point-like X-ray sources with luminosities considerably in excess of the maximum luminosity of a 10 solar mass accreting black hole. X-ray spectra of ULXs have been investigated in many previous studies. However, sparse observations render it more difficult to convincingly distinguish between two competing scenarios, i.e. sub-critical accreting intermediate-mass black hole and super-critical accreting stellar remnant black hole. Here, we report our investigation of a sample of nearby ULXs, as well as their host galaxies, in order to get a more complete understanding of their nature. Multiwavelength analysis was applied to study these ULXs. From X-ray study, we found that most ULXs in our sample do exhibit spectral variability in which higher energy X-ray predominates the spectra as the source becomes brighter. We also compiled some properties of the host galaxies, as a proxy of ULX environment. We found that ULXs in our sample can neither be exclusively associated with a certain type of morphology, nor with high nuclear activity. We also found that 12 out of 19 ULXs reside in host galaxies with star formation rate of less than 1 solar mass per year.

1. Introduction

Ultraluminous X-ray Sources (ULXs) were discovered from X-ray observations of nearby galaxies in the late 1970s by using *Einstein* satellite [1]. They share some general X-ray properties, despite possessing some individual characteristics, one of which is their being bright point-like sources which are not located in the nuclear region of a galaxy. It means that their high luminosity, which is considerably in excess of the maximum lumininosity of $10M_{\odot}$ black hole, cannot be associated with supermassive black hole (SMBH) activities. Currently, there is no clear evidence of the presence of any ULX in our Galaxy. Some X-ray studies have shown that general X-ray properties of ULXs resemble those of Galactic black hole binaries (GBHBs) so that we may in general assume the same energy generation mechanism as that of GBHBs for ULXs, namely accretion flows onto a black hole (see [2] for a review). Mass accretion rate is a key information since it determines the physical processes that happened in the accretion disk which in turns govern the produced radiation, in terms of both the spectral shape and the luminosity. However, it cannot be easily derived from observation. Fortunately, both luminosity and black hole mass can be derived from observation which in turns can be used to constrain mass accretion rate. For a fixed luminosity, higher black hole mass implies lower mass accretion rate, and vice versa. Therefore, ULXs can either be explained with higher black hole mass or higher mass accretion rate.

Content from this work may be used under the terms of the Creative Commons Attribution 3.0 licence. Any further distribution of this work must maintain attribution to the author(s) and the title of the work, journal citation and DOI.
Published under licence by IOP Publishing Ltd

The well-known standard model of accretion disk [3] has been widely used to model soft X-ray thermal spectra of GBHBs in which the disk is assumed to be geometrically thin and optically thick so that the produced radiation in each radius can be approximated with blackbody radiation. The total spectrum can be obtained from the integration of blackbody spectra from all radii of the disk. The standard model has a number of drawbacks when applied to ULXs. Observationally, many ULX spectra deviate from that of the standard model. Those that can be fitted with the standard spectra usually require very high (> 2 keV) or very low (< 0.5 keV) inner disk temperature. Theoretically, the assumptions made for the standard model will eventually break down as one extends the value of mass accretion rate into a higher or lower regime. Different spectral features are expected to emerge due to different radiation processes in the disk as the assumptions of the standard model broke down. Therefore, X-ray spectral study is very important to understand the nature of ULXs. Many previous X-ray spectral studies focused on individual ULXs (e.g. [4],[5],[6]). Several attempts have also been made to study multiple ULXs using high quality, but sparse, *XMM-Newton* data to get insight into the spectral variability of ULXs (e.g. [7],[8]). However, sparse observations render it more difficult to convincingly distinguish between two competing scenarios, i.e. sub-critical accreting intermediate-mass black hole (IMBH) and super-critical accreting stellar remnant black hole. Here, we report our investigation of a sample of nearby ULXs, using long term monitoring data of *Swift* XRT, in order to investigate the general X-ray spectral variability of ULXs in our sample. Previous studies on X-ray spectral variability using *Swift* XRT data have been successful despite focusing on individual ULXs (e.g. [9],[10]). To get a more complete understanding of their nature, multiwavelength study was also conducted to these ULXs. We briefly describe our data and method in section 2 and we present our results and discussion in section 3. We finally conclude our work in section 4.

2. Data and Method

A sample of nearby (< 10 Mpc) ULXs was constructed from a catalog of ULXs [11] which utilized *Chandra* ACIS Survey. This catalog was chosen due to high spatial resolution of *Chandra* X-ray mission. There are 56 ULXs in 36 host galaxies within a distance of 10 Mpc.

Swift XRT data were used to investigate X-ray spectral variability in our sample. We collected all available *Swift* XRT data in the archive since 2006 until July 2016. From the X-ray image, we found that about half of ULXs in our sample require special treatments in data reduction and analysis, which were finally excluded in our current study. These are ULXs which are located very close to the central region of the host galaxy, coincide with any diffuse sources, and/or located very close to other point sources. It reduces the number of ULXs in our sample to 19 ULXs in 17 host galaxies. For each ULX, light curves and spectra were extracted from a circular region with a radius of 47 arcseconds centered at the source position reported in [11]. Another nearby circular region three times a radius of the source region was chosen as the background.

We define hardness ratio as the count rate in 1.5 − 10 keV divided by that of 0.3 − 1.5 keV of background substracted source with a signal-to-noise ratio > 5. Hardness intensity diagram (HID) was then made by plotting the hardness ratio versus the intensity in count/s. Due to short exposure time of *Swift* XRT observation of each ObsID, we did not analyse detail spectrum of each ObsID due to poor statistics. However, there are five ULXs with more frequent observations, compared to the rest, from which we can co-add the spectra and fit them with simple phenomenological models. Before we co-added the spectra, we first groupped observations with the same epoch of redistribution matrix file (RMF). For each RMF group, we further groupped the data based on their count rate to see how the spectra change with intensity. We also checked the hardness ratio of each count rate group to make sure that the co-added spectra have similar hardness ratio.

For the study of host galaxies of ULXs, in our current work we only consider three properties, i.e. morphological classification, stellar mass, and star formation rate (SFR) of the host galaxies. Morphological classification was obtained from various databases (SIMBAD, HyperLEDA, NED and Subarcsecond mid-infrared atlas of local AGN). Stellar mass of a host galaxy was derived from the mass-to-luminosity ratio [12] which requires information of magnitude in B filter and (B-V) color which were collected from the Third Reference Catalogue of Bright Galaxies (RC3). As for the SFR, hydrogen recombination line (H_α) was utilized except for PGC50779 and NGC5128 in which we used total infrared (TIR). Required information to calculate SFR was obtained from Catalog & Atlas of the LV galaxies [13] and *IRAS* archive for H_α and TIR, respectively. The SFR was calculated by following the method described in [14]. We also checked the radial distribution of each ULX within the host galaxy by utilizing information of inclination as well as position angle of the host galaxy to obtain deprojected radial distance of each ULX from the center of each host galaxy.

3. Results and Discussion

ULXs in our sample show variability in intensity with a factor of up to 5 with one exception, PGC23324 X-1, which shows flux variability with a factor of up to 15. For most ULXs we found that the fraction of higher energy photons increases with intensity as can be seen from figure 1 and 2 for more sparse and frequent observations, respectively. Note that we do not plot all ULXs in our sample for the clarity of those figures. PGC23324 X-1 is an example of ULXs in our sample in which soft photons predominate in all available observations. Based on this finding, the spectral energy distribution is expected to vary as a function of intensity. Therefore, we need to analyzed co-added spectra of ULXs in our sample to see how the spectra change with intensity.

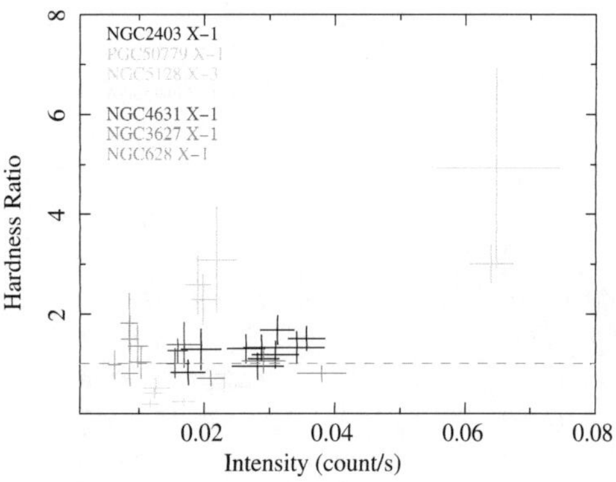

Figure 1. HID for ULXs with sparse observations.

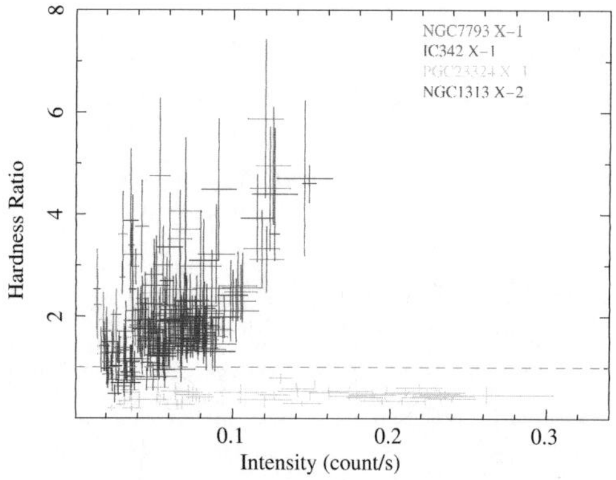

Figure 2. HID for ULXs with frequent observations.

We managed to obtain co-added spectra of five ULXs (NGC7793 X-1, NGC1313 X-2, NGC3031 X-3, IC342 X-1, and PGC23324 X-1). We fitted those spectra with single power law model `PO` and single disk blackbody model `DISKBB` by using `XSPEC`. We used `TBABS` to model Galactic and extragalactic absorption. The Galactic HI column density for each ULX was adopted from [15] while the extragalactic absorption was treated as free parameter. Both single power law and disk blackbody models can in general fit the spectra of these five ULXs. Due to low statictics, especially at higher energy (> 5 keV), we did not attempt to add more sophisticated model, such as thermal comptonization, which are widely suggested to model curvature at higher energy in ULX spectra.

Figure 3 shows the spectra of IC342 X-1 fitted with single power law model. Black and red colors represent spectra of similar count rates but different RMFs. These two spectra are very similar which implies that spectra of similar count rates do not change with time. Red, green, and blue colors represent spectra of the same RMF but of different count rate ranges, which increase from red to blue. The variability of the spectra is minimum at lower energy. The obtained values of photon index (Γ) are rather steep. The evolution of the photon index as a function of unabsorbed luminosity in $0.3 - 10$ keV range can be seen in figure 4 (upper panel). The photon index does not vary much with luminosity despite the apparent change in the spectral shape. The fitting result with single disk blackbody model is shown in figure 4 (lower panel). We can see that the obtained innermost disk temperature (kT_{in}) is rather high (> 1 keV). The change in the spectral shape, together with the obtained values of the inner disk temperature and unabsorbed luminosity, can be interpreted as the evolution from the broadened disk spectra to those of the ultraluminous spectra with curvature at high energy, as the source becomes brighter (see [2] and references there in). The pattern of spectral variability found in this study is consistent with some earlier studies (e.g. [9],[16]). Similar to the evolution of the photon index, the innermost disk temperature does not vary much with luminosity. Similar results were obtained for the other ULXs except NGC7793 X-1, which shows rather flat power law spectra and extremely high innermost disk temperature (> 2 keV).

In a small number of ULXs (not shown in figure 1 and 2), we see an opposite trend in their hardness intensity diagrams, in which the fraction of soft photons increases as the source becomes brighter. Unfortunately, these ULXs were not frequently observed that we could not get decent spectral information.

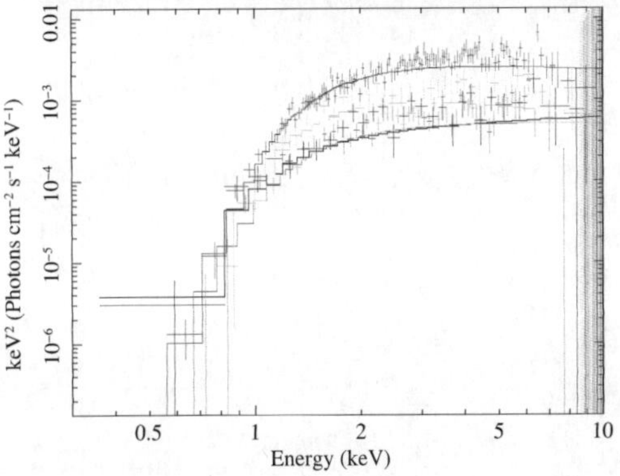

Figure 3. X-ray spectra of IC342 X-1 fitted with single power law model (see text for details).

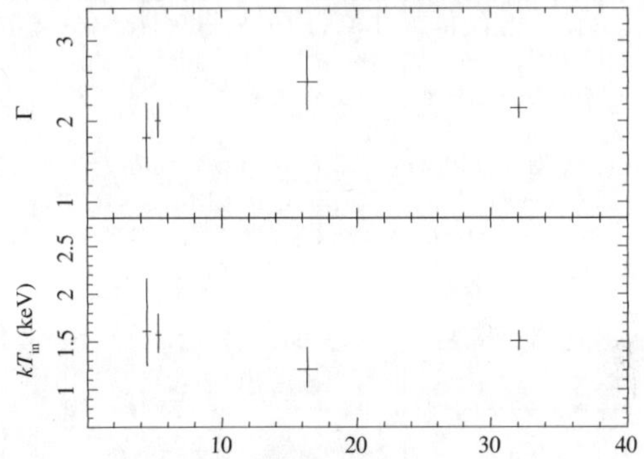

Figure 4. Photon index (Γ) and innermost disk temperature (kT_{in}) of IC342 X-1.

From our study of the three properties of the host galaxies, we found that 37% of the host galaxies in our sample are having peculiar morphology, 26% are categorized as interacting, 37% are found to have starburst activity, and 47% are known as AGNs. ULXs in our sample are located in galaxies with stellar mass in the range of $10^8 - 10^{11} M_\odot$. In terms of SFR, we found that the SFR value of the host galaxies in our sample is rather low, with a range of $0.05 - 2 M_\odot$/yr. In fact, 12 out of 19 ULXs are located in host galaxies with SFR less than $< 1 M_\odot$/yr. We also confirmed that the deprojected position of all ULXs in our sample are located within two R_{25} radii and most of them are located within one R_{25} radii.

Based on the result of morphological classification, abnormal activities in the host galaxies are not a requirement for the existence of a ULX in a galaxy. The fact that we have not found any ULX in our own Galaxy remains puzzling. However, some ULXs are reported to be transient.

4. Conclusion

Based on our multiwavelength study, it appears that inhomogeneity exist even in a small sample of nearby ULXs, both in X-ray and host galaxy properties. However, majority of ULXs in the sample exhibit similar trend in X-ray spectral variability, in which the fraction of high energy photons increases with intensity. Based on the available spectra, we either obtained rather hot thermal spectra or rather steep non-thermal spectra, both of which are difficult to explain by using low mass accretion rate model.

Acknowledgments

This work made use of data supplied by the UK Swift Science Data Centre at the University of Leicester. This work is supported by P3MI ITB grant 2018. KV thanks AHEAD Trans-national access to X-ray data analysis program cycle 2. We thank anonymous reviewers for their careful reading and helpful suggestions. We also thank Sulistiyowati for her contribution in developing the automation system for Swift data reduction and extraction in our institution.

References

[1] Fabbiano G 1989 *ARA&A* **27** 87
[2] Kaaret P, Feng H and Roberts T P 2017 *ARA&A* **55** 303
[3] Shakura N I and Sunyaev R A 1973 *A&A* **24** 337
[4] Feng H and Kaaret P 2006 *ApJL* **650** 75
[5] Kajava J J E, Poutanen J, Farrell S A, Grisé F and Kaaret P 2012 *MNRAS* **442** 990
[6] Yoshida T, Isobe N, Mineshige S, Kubota A, Mizuno T and Saitou K 2013 *PASJ* **65** 14
[7] Stobbart A M, Roberts T P and Wilms J 2006 *MNRAS* **368** 397
[8] Gladstone J C, Roberts T P and Done C 2009 *MNRAS* **397** 1836
[9] Vierdayanti K, Done C, Roberts T P and Mineshige S 2010 *MNRAS* **403** 1206
[10] La Parola V, D'Aí A, Cusumano G and Mineo T 2015 *A&A* **580** 6
[11] Liu J 2011 *ApJS* **192** 10
[12] Bell E F, McIntosh D H, Katz N and Weinberg M D 2003 *ApJS* **149** 289
[13] Karachentsev I D and Kaisina E I 2013 *AJ* **146** 44
[14] Kennicutt R C Jr and Evans N J 2012 *ARA&A* **50** 531
[15] Kalberla P M W, Burton W B, Hartmann D, Arnal E M, Bajaja E, Morras R and Pöppel W G L 2005 *A&A* **440** 775
[16] Luangtip W, Roberts T and Done C 2016 *MNRAS* **460** 4417

Mass determination of 5 relaxed galaxy clusters with Chandra X-ray data

H Wulandari, A H Fikri, K Vierdayanti, A N I Putri and D G Ramadhan

Astronomy Research Division, Faculty of Mathematics and Natural Sciences, Institut Teknologi Bandung, Jl. Ganesha 10 Bandung 40132

E-mail: hesti@as.itb.ac.id

Abstract.
We report masses of 5 relaxed galaxy clusters (Abell 2204, Abell 478, RXJ 1532.9+3021, Zwicky 2089, and Abell 2597) obtained from analysis of Chandra X-ray satellite ACIS data. Assuming that the clusters are in hydrostatic equilibrium, masses can be estimated from temperature and density profiles of the clusters. In order to produce temperature and density profiles we performed data reduction, spectrum extraction, fitting of multiple observational spectra, and deprojection procedures. Density and temperature profiles were fitted with a single beta model and an exponential model respectively. We determined masses of the clusters enclosed by R_{2500}, which is the radius where the density equals to 2500 times the critical density of the universe. We compare our results with other mass determinations employing different methods, i.e. Sunyaev Zeldovich Effect (SZE) and lensing, where available and found that within uncertainties they are in a good agreement.

1. Introduction

Clusters of galaxies are the most massive structures in the universe that have reached equilibrium. The gravitationally bound systems consist of hundreds to thousands galaxies, have diameters of $2 - 10$ Mpc and a typical range of mass between $10^{13} - 10^{15} M_{\odot}$. Clusters are not only unique astrophysical laboratories in which many astrophysical phenomena occur, but also are powerful cosmological probes. A massive cluster is large enough to be considered as a representative volume of the Universe. The matter contents of the most massive clusters represent the matter content of the universe. Clusters therefore are commonly used in constraining cosmological parameters, and in studies concerning dark matter, dark energy, as well as large scale structures.

Cluster mass is a physical yet not observable parameter that should be estimated for cosmological studies. There are several methods for cluster mass determination. We can measure velocity dispersions of galaxies in a cluster and derived its mass from the virial theorem. This was done for example by Zwicky in 1933, who found that mass in the form of galaxies in Comma cluster is much less than the total mass derived from dynamics. He was the first to coin the term "dark matter" for the excess mass needed to explain the discrepancy. In a gravitational lensing event by a cluster, the light rays from background galaxies are bent towards us. By studying distortions in the images of the background galaxies we can estimate the total mass of the cluster. Observing a cluster in X-ray and assuming that it is in hydrostatic equilibrium,

Content from this work may be used under the terms of the Creative Commons Attribution 3.0 licence. Any further distribution of this work must maintain attribution to the author(s) and the title of the work, journal citation and DOI.
Published under licence by IOP Publishing Ltd

Journal of Physics: Conference Series **1354** (2019) 012014 doi:10.1088/1742-6596/1354/1/012014

the total mass of the cluster can be estimated from its temperature and density profiles. This is the method that we employed for the five clusters we report here. The last method is the so called Sunyaev Z'eldovich effect (SZE). This effect occurs when cosmic microwave background (CMB) photons are inverse Compton scattered to higher energies by electrons in Intra Cluster Medium (ICM), causing a shift and distortion in the shape of the CMB spectrum. The size of the wavelength shift is dependent on the mass of the galaxy cluster the photons passed through.

2. Data and Method

In this work we analyzed data of five clusters of galaxies (see Table 1) observed by Chandra X-ray satellite with ACIS instruments without any gratings. These clusters are identified as relaxed by [1] based on their morphologies in X-ray imaging data. To perform data reduction we used Chandra analysis tool CIAO version 4.9 and Chandra calibration CALDB version 4.7.3. Compared to optical observations, Xray-observations are photon-starved. To improve statistics we therefore combined multiple observations data for each cluster, whose ObsIDs are listed in Table 1.

Table 1. Clusters data

Obj Name	RA	Dec	Redshift	ObsID
Abell 2204	16 : 32 : 46.920	05 : 34 : 32.86	0.1524	499, 6104, 7940
Abell 478	04 : 13 : 25.345	10 : 27 : 55.15	0.0883	1669, 6102
RXJ 1532.9+3021	15 : 32 : 53.781	30 : 20 : 58.72	0.345	1649, 1665, 14009
Zwicky 2089	09 : 52 : 49.183	51 : 53 : 05.27	0.235	7897, 10463
Abell 2597	23 : 35 : 19.779	−12 : 07 : 27.63	0.0854	922, 6934, 7329

All data were reduced using standard data reduction processes. We started with dirty images (level 1 or L1), then performed correction for bad pixels and bad columns of the CCDs to get L2 images. Since clusters are extended objects, point sources should be removed and we did this with `wavdetect` tools of Chandra. We also removed flares coming from interactions of high energy particles ($9.5 - 12$ keV) with the CCDs, renormalized the blank sky, and then did background subtraction to the to get clean images.

Since all the clusters in our study are relaxed, we assumed that the temperatures of the clusters are spherically symmetric structured and applied a deprojection technique. For each cluster, we first combined the clean images of all ObsIDs, then set annuli based on the net counts of the combined images. We fixed a minimum number of 7 annuli for each cluster. We simultaneously fitted the spectra of multiple ObsIDs while performing the deprojection process. All spectra were fitted with Astrophysical Plasma Emission Code (APEC) model for the hot plasma emission (http://cxc.harvard.edu/ciao/dictionary/apec.html) convoluted with photoelectric absorption with Wisconsin cross-section (WABS) model [2] for the Galactic absorption,

$$SpectrumModel = WABS(n_H) * APEC(T, z, Z, norm) \tag{1}$$

where n_H, T, z, Z and $norm$ respectively denote Galactic hydrogen column density, gas temperature, redshift, metalicity and normalization. The last is dependent on electron number density n_e and cosmological model,

$$norm = \frac{10^{-14}}{4\pi\, D_A(1+z)^2} \int n_e n_H dV \tag{2}$$

where D_A is the angular diameter distance to the object, which depends on cosmological model. We used a standard ΛCDM universe and adopted redshift and metallicity data from Archive of Chandra Cluster Entropy Profile Tables (ACCEPT) available at (https://web.pa.msu.edu/astro/MC2/accept/). The Galactic hydrogen column densities in the directions of the objects under study were obtained from Galactic N_H calculator (http://www.swift.ac.uk/analysis/nhtot/) provided by SWIFT X-ray mission.

3. Results and Discussions

The deprojected electron density profiles are shown in Figure 1 for two of the five clusters under study. We fitted the profiles with a single-β model [3],

$$n_e(r) = n_{e0}(1 + (r/r_c)^2)^{-3\beta/2} \tag{3}$$

where central electron number density n_{e0}, core radius r_c and β are free parameters to be obtained from the fitting. We see that the profiles are very well fitted with the model.

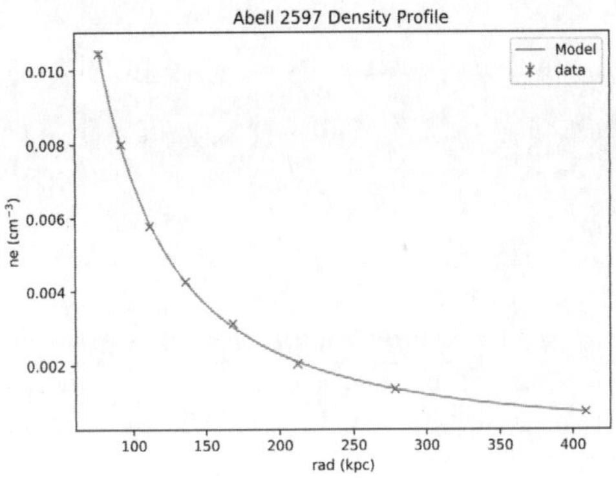

Figure 1. Electron density profile of Abell 2597 fitted with a single β model.

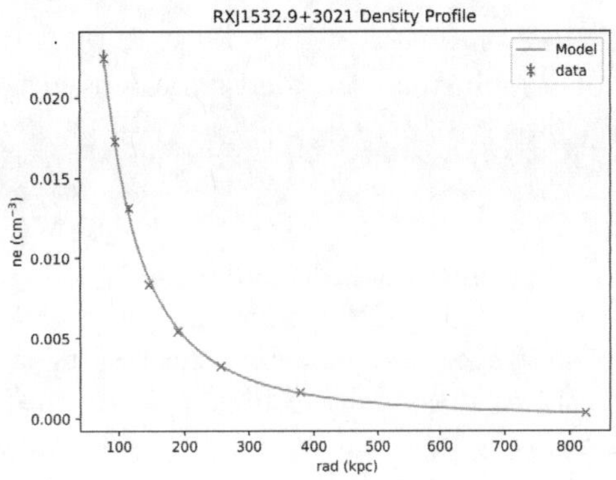

Figure 2. Electron density profile of RXJ1532.9-3021 fitted with a single β model.

In Figure 2 the deprojected temperature profiles for the two clusters are shown and fitted with an exponential model [3],

$$T(r) = T_0\, exp(-(r-b)^2/c^2) \tag{4}$$

where T_0 is the central temperature in keV, b and c are normalization radii. We see that the temperature decreases towards the center, which may be ascribed to the gas cooling whose rate is proportional to the electron number density.

Assuming that the clusters are spherically symmetric and are in hydrostatic equilibrium, the total mass of the clusters within radius r can be determined from the radial profiles of the gas density and temperature as follows

$$M(<r) = -\frac{kT(r)r}{G\mu m_p}\left(\frac{r}{\rho_{gas}}\frac{d\rho_{gas}}{dr} + \frac{r}{T}\frac{dT}{dr}\right) \tag{5}$$

where k_B, G, m_p and μ are the Boltzmann constant, the gravitational constant, the proton mass and the mean molecular weight of the gas in units of m_p respectively. The mean molecular weight of a fully ionized gas with a standard cosmic abundance is $\mu = 0.6$ [3]. The gas mass densities

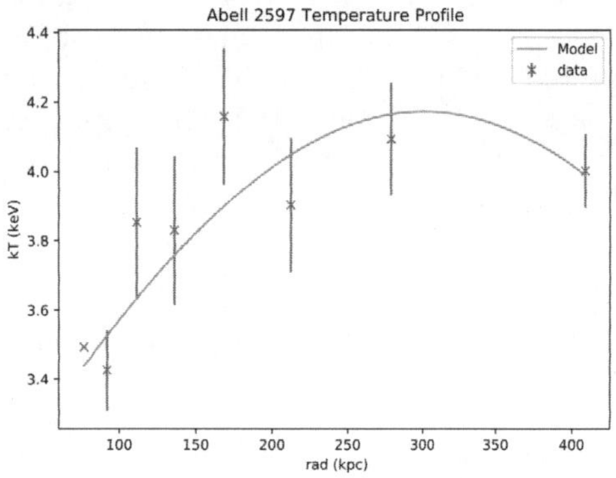

Figure 3. Temperature profile of Abell 2597 fitted with an exponential model.

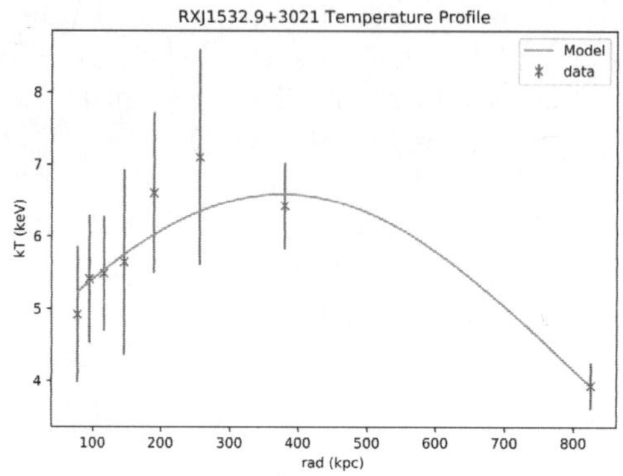

Figure 4. Temperature profile of RXJ1532.9-3021 fitted with an exponential model

ρ_{gas} of the clusters were determined from the electron number densities that were previously obtained,

$$\rho_{gas}(r) = n_e(r)\ \mu\ m_p \tag{6}$$

For each cluster we estimated the total mass enclosed by R_{2500}, which is the radius where the density is equal to 2500 times the critical density ρ_c of the universe. We adopted R_{2500} values from [4]. Table 1 summarizes the results of our mass determinations for all five clusters. As comparisons masses determined by other authors employing different methods are shown, i.e. from SZE [4] and weak lensing [5]. The weak lensing results are shown for Hubble parameter $h = 0.7$, the same value we used in our mass determinations.

Table 2. Galaxy clusters masses determined from different methods

Obj Name	R_{2500} (kpc) [4]	M_{2500} (This work) X-ray	M_{2500} [4] SZE	M_{2500} [5] Lensing
Abell 2204	830 ± 79	5.6 ± 1.0	9.20 ± 1.70	$4.73^{+0.68}_{-0.67}$
Abell 478	570 ± 80	2.0 ± 0.5	2.80 ± 1.10	—
RXJ 1532.9+3021	499 ± 107	1.28 ± 0.06	2.90 ± 1.60	$1.81^{+0.68}_{-0.67}$
Zwicky 2089	367 ± 72	0.47 ± 0.08	0.83 ± 0.30	—
Abell 2597	268 ± 21	0.294 ± 0.25	0.288 ± 0.047	—

In general our results are in agreement within uncertainties with the other two mass determination methods. For Abell 2204 our result is consistent with that of lensing, but SZE method gives considerably higher mass. It should be noted here that [4] assumed isothermal clusters, and used only density profiles in their mass determination.

It is often discussed that analysis of lensing clusters may be affected by orientation biases, while the assumption of hydrostatic equilibrium in X-ray mass determination method may not always be accurate. But since here we are dealing with relaxed clusters, the two concerns might be considered not worrisome and the mass estimates here are reasonable.

4. Conclusions

By assuming hydrostatic equilibrium and employing deprojection technique, we have determined masses of five clusters of galaxies observed by Chandra X-ray mission. In general our results show within uncertainties good agreement with results of SZE and weak lensing methods, i.e. the three methods are consistent with each other. Lensing events can only occur in particular configuration, whereas X-ray observations are in principle doable for any clusters. SZE needs additional observations in radio waves beside X-ray observations. Therefore, X-ray cluster mass determination offers an expeditious way if we need to determine masses of a large number of clusters, which is common in cosmological studies.

Acknowledgments

The authors thank ITB for supporting this work through P3MI research grant 2017 and 2018.

References

[1] Mantz A B, Allen S W, Morris R G, Schmidt R W, von der Linden A, and Urban O 2015 *MNRAS* **449(1)** 199
[2] Morrison R and McCammon D 1983 *ApJ* **270** 119
[3] Schneider P 2006 *Extragalactic Astronomy and Cosmology* (Heidelberg: Springer-Verlag)
[4] Comis B, De Petris M, Conte A, Lamagna L and De Gregori S 2011 *MNRAS* **418(2)** 1089
[5] Okabe N and Smith G P 2016 *MNRAS* **461(4)** 3794

Journal of Physics: Conference Series **1354** (2019) 012015 doi:10.1088/1742-6596/1354/1/012015

Background Dependence of Pole Position in Kaon Photoproduction

S. Clymton and T. Mart

Departemen Fisika, FMIPA, Universitas Indonesia, Depok 16424, Indonesia

E-mail: samsonclymton@gmail.com, terry.mart@sci.ui.ac.id

Abstract. Pole position has become a new important parameter to describe the properties of nucleon resonance. Compared to the Breit-Wigner parameters, the pole position is less model dependent. The shape of resonance based on our previous formalism is strongly affected by the hadronic form factors. We study the effect of different background parameters on the extracted pole position. We have also used different hadronic form factors to see the effect on the background dependence of pole position. We obtained the necessary parameters by fitting to about 9000 experimental data points. We expect to find appropriate hadronic form factors that lead to the background independence. Therefore, this study will shed a light on the choice of the appropriate hadronic form factors.

1. Introduction

In recent years, the Particle Data Group (PDG) has been prioritizing the pole position over the Breit-Wigner parameters. It is shown in the particle listing of PDG that the pole position of a resonance is written before the Breit-Wigner parameter. Theoretically, pole position is less model dependent compared to the Breit-Wigner paramaterization [1, 2]. Therefore, it provides a good benchmark in many theoretical model that gives resonances position. However, in practice, it could depend on the background by various reasons. Furthermore, there has been a study showing that both properties are needed to describe the resonances [3]. On the other hand, the formalism of consistent interaction using the Lagrangian effective theory to evaluate the scattering amplitude for high spin resonances in kaon photoproduction has been succsesfully constructed and proven to be better than other models [4, 5]. However, there are some problems rising due to the strong momentum dependence in the scattering amplitude. Two of these problems are the false position of resonances and the divergences of amplitude at higher energies. This defect does not occur in our recent study because it is concealed by the background and it is suppressed by the hadronic form factor. Vrancx *et al.* [6] tried to attack this defect by using the suitable hadronic form factor that leads to the correct resonances position and convergence at high energies. This is the reason to study the background independence of pole position of the nucleon resonances in the $\gamma p \to K^+\Lambda$ reaction by using various hadronic form factors.

We use our recent model in kaon photoproduction [5, 7] as the benchmark model to investigate the effect of diferent form factors and variation of the background. The corresponding Feynman diagrams used in this study are shown in Figure 1. This model involves nucleon resonances with spins up to 9/2 and the spin-1/2 and -3/2 hyperon resonances as the background. The background of course varies by the involvement of spin-3/2 hyperons. Furthermore, we use three

Content from this work may be used under the terms of the Creative Commons Attribution 3.0 licence. Any further distribution of this work must maintain attribution to the author(s) and the title of the work, journal citation and DOI.
Published under licence by IOP Publishing Ltd

Journal of Physics: Conference Series **1354** (2019) 012015 doi:10.1088/1742-6596/1354/1/012015

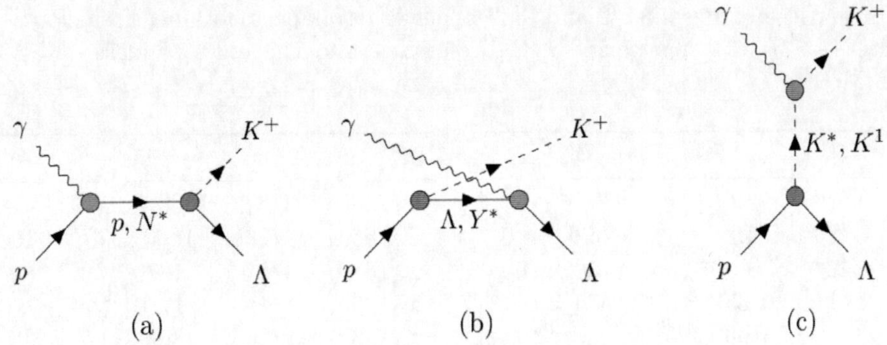

Figure 1. Feynman diagrams for kaon photoproduction $\gamma + p \to K^+ + \Lambda$. Each diagram describes the (a) s-channel, (b) u-channel and (c) t-channel.

different form factors. Each model is fitted to around 9000 experimental data. The extracted Breit-Wigner parameter is used to evaluate the pole position. The effect of certain resonances will be presented by the plots of total cross section.

2. The Pole Position

In principle, the pole position is evaluated at the zero value of denominator in the scattering amplitude. In this way, the scattering amplitude of a resonance will be very large near the pole position. As a consequence, the background amplitude will not affect the property of resonances evaluated at the pole position. Thus, the projection of experimental data to the resonance region must have certain value of pole position that is not affected by the background.

The pole position is defined by

$$\sqrt{s_R} = W_R = M_{\rm pole} - i\Gamma_{\rm pole}/2 \ , \tag{1}$$

which can be easily obtained by imposing the denominator of scattering amplitude to vanish, i.e.,

$$s_R - m_R^2 + im_R\Gamma(s_R) = 0 \ . \tag{2}$$

In this work we use the energy dependent width [8] that is directly proportional to the total width Γ_R. As a consequence, we cannot solve this equation analitically. Only the numerical calculation works in this case. The initial values of m_R and Γ_R are obtained from the particle listing of the Particle Data Book. These values are actually the Breit-Wigner parameters that were used as free parameters in the fitting process, where the upper and lower limits are bounded to the PDG values. The obtained parameters are used to calculate the pole positions of resonances.

3. The Hadronic Form Factor

The problem of divergent scattering amplitude of the resonances in high energy region is an old issue. The well known solution is the introduction of hadronic form factors to suppress this divergence. The various form factors were constructed mainly by specific purpose and sometimes the formula becomes trivial. The most common form factor widely used is the dipole form factor. In our previous study, we used the dipole hadronic form factor for all resonances that reads

$$F_{\rm dip}(s,\Lambda) = \frac{\Lambda^4}{(s-m_R^2)^2+\Lambda^4} \ , \tag{3}$$

Table 1. Properties of the spin-1/2 and -3/2 hyperon resonances obtained from Particle Data Book [1]. Here J^P is the spin-parity of the resonance and the units of mass and width are expressed in MeV.

Y^*	J^P	Mass	Width
$\Lambda(1405)$	$1/2^-$	$1405^{+1.3}_{-1.0}$	50.5 ± 2
$\Lambda(1520)$	$3/2^-$	1519.5 ± 1.0	15.6 ± 1.0
$\Lambda(1600)$	$1/2^+$	1600^{+100}_{-40}	150 ± 100
$\Lambda(1670)$	$1/2^-$	1670 ± 10	35 ± 15
$\Lambda(1690)$	$3/2^-$	1690 ± 5	60 ± 10
$\Lambda(1800)$	$1/2^-$	1800^{+50}_{-80}	300 ± 100
$\Lambda(1810)$	$1/2^+$	1810^{+40}_{-60}	150 ± 100
$\Lambda(1890)$	$3/2^+$	1890^{+20}_{-40}	100^{+100}_{-40}

Y^*	J^P	Mass	Width
$\Sigma(1385)$	$3/2^+$	1383.7 ± 1.0	36 ± 5
$\Sigma(1660)$	$1/2^+$	1660 ± 30	100^{+100}_{-60}
$\Sigma(1670)$	$3/2^-$	1670^{+15}_{-5}	60 ± 20
$\Sigma(1750)$	$1/2^-$	1750^{+50}_{-20}	90^{+70}_{-30}
$\Sigma(1880)$	$1/2^+$	1821 ± 17	300 ± 59
$\Sigma(1940)$	$3/2^-$	1940^{+10}_{-40}	220^{+80}_{-70}
$\Sigma(2080)$	$3/2^+$	2091 ± 7	186 ± 48

where Λ is the cutoff parameter and s is the square of total c.m. energy. Unfortunately in our recent formalism this form factor is too weak to suppress the strong momentum dependence of high spin resonances amplitude. However, as mentioned before, this problem can be also suppressed by the background. However, suppresing the divergence in this way could lead to the background dependence of pole position, since the resonances-like structure is obtained with the help of the background.

Our formalism of high spin resonances interaction is based on the Pascalutsa [9] and Vrancx *et al.* [6] works. In their paper, Vrancx *et al.* discuss the suitable form factor for this formalism. They claim that it can be preserve the position of resonances in terms of their mass and width. The form factor is called multidipole-Gauss as it is made by combining the multidipole and Gaussian form factors. It will be seen that it is actually not the multidipole, but rather the modified form factor that preserves the interpretation of decay width as the FWHM of the resonance. The form factors reads

$$F_{\rm mmG}(s,\Lambda) = \left\{ \frac{m_R^2\tilde{\Gamma}_R^2}{(s-m_R^2)^2 + m_R^2\tilde{\Gamma}_R^2} \right\}^{J-1/2} \exp\left\{ -\frac{(s-m_R^2)^2}{\Lambda^4} \right\} \quad (4)$$

where J is the spin of resonance and $\tilde{\Gamma}_R$ is the modified width that reads as

$$\tilde{\Gamma}_R = \frac{\Gamma_R}{\sqrt{2^{1/(2J)} - 1}} \; . \quad (5)$$

This form factor can be modified by subtituting $m_R^2\tilde{\Gamma}_R^2$ with the cutoff parameter Λ to obtain the actual multidipole-Gauss one. Therefore, along with those two kind of form factors we try to test this form factor in our recent model of kaon photoproduction with two variations of the background.

4. Results and discussion

The hyperon resonances (Y^*) used in the background can be seen in Table 1. There is two variations of the background used in this work, i.e., the model without the spin-3/2 hyperon resonances and that with the spin-3/2 hyperon resonances. Along with this, we fitted 3 kinds of hadronic form factors to test the background independence of resonances pole position. The obtained pole positions compared to the data from PDG are shown in Table 2. From this Table we can clearly see that in all models the resonances $N(1520)$, $N(1675)$, $N(1680)$ and $N(1720)$ are consistent with the PDG. There are a number of resonances which have large discrepancies with the PDG values, but consistent in all models, i.e., the $N(1880)$, $N(1990)$, $N(2000)$, $N(2060)$ and $N(2220)$. The result can be used to evaluate the actual pole position of the resonances or for further study of the resonance role in our model.

In this work our concern is not to compare our result with the PDG, but instead we want to see the background dependence of pole position. Thus, we try to calculate the absolute value of of pole position difference of the two different backgrounds, in each of hadronic form factors, i.e.,

$$|\Delta W_R/W_R| = \sqrt{\frac{[m_R(1) - m_R(2)]^2 + [\Gamma_R(1) - \Gamma_R(2)]^2/4}{m_R^2(1) + \Gamma_R^2(1)/4}}\,, \tag{6}$$

where the notations (1) and (2) refer to the models without and with spin-3/2 hyperon resonances, respectively. Table 3 presents the result of numerical calculation. In general, we found that models using multidipole-Gauss form factor are more independent than those using other two form factors. The use of modified multidipole-Gauss form factors leads to the worst result. By looking at Table 3 one can say that in general the result is consistent for all models because the largest discrepancy is only 4.4%. However, this is not completely true, since we have just involved the spin-3/2 hyperon resonances. Furthermore, the inclusion one nuleon resonance might become a background for another nucleon resonance. Thus, we trivially set the accepted difference to be 1%. As a consequence, based on the Table 3, in the model with modified multidipole-Gauss form factor more nucleon resonances are independent to the change of background compared to dipole one. Interestingly, the $N(1535)$ resonance has the difference value larger than 1% for all form factor models. This indicates that the $N(1535)$ is probably

Table 2. Comparison of the $M_{\rm pole}$ and $\Gamma_{\rm pole}$ obtained from Particle Data Book [1] and those obtained from calculation of the extracted mass and width of 6 different models with 3 different form factors and 2 variations of background. The notations (1) and (2) in the column title refer to the model without and with spin-3/2 hyperon resonances, respectively.

Resonans	PDG		$F_{\rm dip}$ (1)		$F_{\rm dip}$ (2)		$F_{\rm mmG}$ (1)		$F_{\rm mmG}$ (2)		$F_{\rm mG}$ (1)		$F_{\rm mG}$ (2)	
	$M_{\rm pole}$	$\Gamma_{\rm pole}$	$M_{\rm pole}$	$\Gamma_{\rm pole}$	$M_{\rm pole}$	$\Gamma_{\rm pole}$	$M_{\rm pole}$	$\Gamma_{\rm pole}$	$M_{\rm pole}$	$\Gamma_{\rm pole}$	$M_{\rm pole}$	$\Gamma_{\rm pole}$	$M_{\rm pole}$	$\Gamma_{\rm pole}$
N(1440)	1369	189	1305	174	1305	174	1382	177	1344	160	1305	174	1305	174
N(1520)	1510	110	1495	102	1487	115	1477	113	1485	100	1487	115	1487	115
N(1535)	1510	170	1508	169	1475	163	1430	209	1474	225	1474	225	1475	163
N(1650)	1655	135	1604	261	1613	232	1653	175	1639	183	1627	171	1627	171
N(1675)	1660	135	1640	148	1640	148	1627	166	1627	166	1640	148	1640	148
N(1680)	1675	120	1651	123	1652	123	1645	132	1645	132	1635	130	1642	122
N(1700)	1700	200	1670	129	1692	152	1687	153	1693	153	1598	154	1584	169
N(1710)	1720	230	1658	175	1658	175	1689	64	1689	65	1710	75	1710	75
N(1720)	1675	250	1643	204	1652	188	1639	208	1639	207	1654	186	1658	188
N(1860)	1830	250	1856	221	1862	217	1851	225	1844	229	1793	218	1787	228
N(1875)	1875	200	1714	244	1787	277	1779	252	1771	269	1740	255	1724	248
N(1880)	1870	220	1786	298	1786	298	1770	285	1766	282	1786	298	1786	298
N(1895)	1907	100	1879	208	1876	225	1873	245	1872	251	1875	230	1875	233
N(1900)	1920	215	1867	253	1862	261	1824	307	1820	311	1867	253	1867	253
N(1990)	2030	240	1866	241	1875	237	1908	279	1841	252	1900	220	1900	220
N(2000)	2030	380	1933	274	1933	274	1859	265	1902	283	1904	284	1904	284
N(2060)	2030	400	1857	341	1857	341	1874	335	1858	325	1836	329	1836	329
N(2120)	2115	345	1925	349	1885	355	1911	341	1911	341	1946	361	1951	363
N(2190)	2075	450	2046	252	2027	245	2028	249	2045	252	1965	227	1970	233
N(2220)	2170	480	2020	228	2020	228	2031	226	2048	221	2048	221	2048	221
N(2250)	2200	450	2060	274	2085	266	2087	299	2089	298	2138	285	2120	291

Table 3. Comparison of the absolute value of the nucleon resonance pole position diference between two different background models for each of the hadronic form factors ($|\Delta W_R/W_R|$). The result is given in percentage (%).

Resonans	J^P	F_{dip}	F_{mmG}	F_{mG}	Resonans	J^P	F_{dip}	F_{mmG}	F_{mG}
N(1440)	$1/2^+$	0	2.8	0	N(1880)	$1/2^+$	0	0.2	0
N(1520)	$3/2^-$	0.7	0.7	0	N(1895)	$1/2^-$	0.5	0.2	0.1
N(1535)	$1/2^-$	2.2	3.1	2.1	N(1900)	$3/2^+$	0.3	0.2	0
N(1650)	$1/2^-$	1.1	0.9	0	N(1990)	$7/2^+$	0.5	3.6	0
N(1675)	$5/2^-$	0	0	0	N(2000)	$5/2^+$	0	2.4	0
N(1680)	$5/2^+$	0.1	0	0.5	N(2060)	$5/2^-$	0	0.9	0
N(1700)	$3/2^-$	1.5	0.4	1	N(2120)	$3/2^-$	2.1	0	0.3
N(1710)	$1/2^+$	0	0	0	N(2190)	$7/2^-$	0.9	0.8	0.3
N(1720)	$3/2^+$	0.7	0	0.2	N(2220)	$9/2^+$	0	0.8	0
N(1860)	$5/2^+$	0.3	0.4	0.4	N(2250)	$9/2^-$	1.2	0.1	0.9
N(1875)	$3/2^-$	4.4	0.7	0.9					

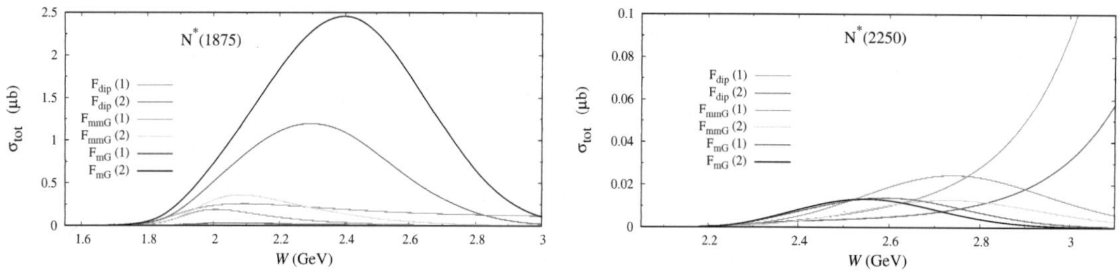

Figure 2. Comparison of the total cross section of $N(1875)$ (left) and $N(2250)$ (right) resonances for models with three different hadronic form factors and two variations of background.

not so important in this reaction.

Figure 2 shows the total cross section for nucleon resonances $N(1875)$ and $N(2250)$, whereas Table 3 for these resonances the modified multidipole-Gauss form factor has the lowest difference. Figure 2 confirms this. For the two resonances the model using this form factor has similar plots, but different in strength. On the other hand, the cross section of $N(1875)$ with multidipole-Gauss form factor has a similar pattern with the modified one. The worst result come from the dipole form factors. For the $N(2250)$ resonance we observe that a divergence in high energy region, but addition of hyperon resonances could suppress it. The similar case is also found for the $N(1875)$ resonance, but the divergence is relatively soft due to the proportionality of momentum dependence to the spin. Furthermore, the result obtained after the inclusion of spin-3/2 hyperon resonances leads to a better resonance-like structure, but it shifts the pole position from its original one, as we can see in Table 3.

5. Summary and conclusion

We have studied the background dependence of pole position of resonances in kaon photoproduction by using 3 different hadronic form factors. The result shows that the multidipole-Gauss form factor is the best for achieving the background independent pole position. However, there is still the lack of knowledge of the source of this problem. Therefore, further study to elaborate the source of this problem and to build a suitable form factor, that can improve the model, is strongly advised.

Acknowledgments
This work has been partly supported by the PITTA Grant Universitas Indonesia No. 2320/UN2.R3.1/HKP.05.00/2018.

References
[1] Patrignani C *et al* [Particle Data Group] 2014 *Chin. Phys.* C **40** 100001
[2] Anisovich A V *et al* 2012 *Eur. Phys. J.* A **48** 15
[3] Ceci S *et al* 2017 *Sci. Rep.* **7** 45246
[4] Mart T, Clymton S and Arifi A J 2015 *Phys. Rev.* D **92** 094019
[5] Clymton S and Mart T 2017 *Phys. Rev.* D **96** 054004
[6] Vrancx T *et. al* 2011 *Phys. Rev.* C **84** 045201
[7] Clymton S 2018 Thesis Universitas Indonesia
[8] Lee F X *et. al* 2001 *Nucl. Phys.* A **695** 237
[9] Pascalutsa V 1998 *Phys. Rev.* D **58** 096002

Journal of Physics: Conference Series **1354** (2019) 012016 doi:10.1088/1742-6596/1354/1/012016

Low Energy Theorem for Kaon Photoproduction

S. Sakinah, S. Clymton and T. Mart
Departemen Fisika, FMIPA, Universitas Indonesia, Depok 16424, Indonesia
E-mail: terry.mart@sci.ui.ac.id

Abstract. We study the Low Energy Theorem (LET) for kaon photoproduction at threshold region. To this end we expand the transition amplitude in terms of the kaon mass and the averaged mass of nucleon and Λ-hyperon. The background amplitudes are obtained from the appropriate Feynman diagrams. The amplitudes have been so far expanded up to the 3th order.

1. Introduction

Low Energy Theorem (LET) has been succesfully studied in the case of pion photoproduction at the threshold region [1]. The expansion of the transition amplitude was performed by means of the pseudo-vector theory in terms of the ratio between pion and nucleon masses. The theorem was first put forward by Kroll and Ruderman [3] and is based on the assumption that the pion mass is significantly smaller than the nucleon mass as well as on the hypothesis of the partial conservation of axial current [2]. The three-momentum of pion at the threshold is found to be small. The photon and pion four-momenta have been considered to be small.

For kaon photoproduction LET has not been considered before, since kaon mass is clearly much heavier. Therefore, in this paper we propose to derive LET for the threshold cross-section of kaon photoproduction. The cross-section will be derived by expanding the transition amplitude of kaon photoproduction in the pseudovector theory. The amplitudes are extracted from the appropriate Feynman diagrams. The background amplitude in the pseudo-vector theory is given in Ref. [4].

As widely known, kaon photoproduction uses the nucleon as a target and produces a Λ-hyperon as a recoil. Since the nucleon and Λ-hyperon have different masses, we propose to expand the cross-section of kaon photoproduction in terms of the ratio between kaon mass and the averaged mass of nucleon and Λ. Besides that, we already know that the mass of kaon is heavier than the mass of pion and, furthermore, the threshold of kaon photoproduction is significantly higher than that of the pion. Clearly, we need to expand the amplitude to the much higher order than in the case of pion. In this work, we perform the expansion up to the 3th order.

2. FORMALISM AND METHOD

The Feynman diagrams for kaon photoproduction are shown in Fig. 1. The background amplitude is constructed by using these Feynman diagrams in the pseudo-vector theory [4], i.e.,

$$\mathcal{M}_B = i\frac{g_{K\Lambda N}}{(m_\Lambda + m_p)}\bar{u}_\Lambda \left[\gamma_5 \not{q}\left\{\frac{\not{p}_p + \not{k} + m_p}{s - m_p^2}(\not{\epsilon} e F_1^p + i\sigma^{\mu\nu}\epsilon_\mu k_\nu \mu_p F_2^p) - \frac{k.\epsilon}{k^2}eF_1^p\right\}\right. \quad (1)$$

Content from this work may be used under the terms of the Creative Commons Attribution 3.0 licence. Any further distribution of this work must maintain attribution to the author(s) and the title of the work, journal citation and DOI.
Published under licence by IOP Publishing Ltd

Journal of Physics: Conference Series 1354 (2019) 012016 doi:10.1088/1742-6596/1354/1/012016

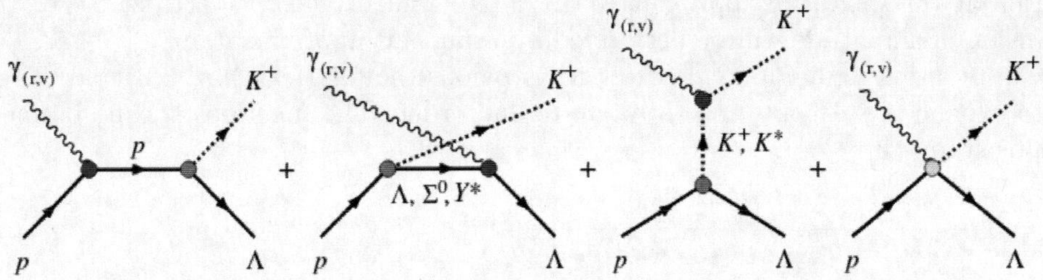

Figure 1. Feynman diagrams of kaon photoproduction for the s-, t-, and u-channel, along with the contact diagram to preserve the gauge invariance. [4]

$$+i\sigma^{\mu\nu}\epsilon_\mu k_\nu \mu_\Lambda F_2^\Lambda \frac{\not{p}_\Lambda - \not{k} + m_\Lambda}{u - m_\Lambda^2}\gamma_5 \not{q} + \gamma_5(\not{q} - \not{k})\left\{\frac{2q - k}{t - m_k^2} + \frac{k.\epsilon}{k^2}\right\} eF^K \Bigg] u_p$$

$$+i\frac{g_{K\Sigma N}}{(m_\Lambda + m_p)}\bar{u}_\Lambda \left[i\sigma^{\mu\nu}\epsilon_\mu k_\nu \mu_T F_2^T \frac{\not{p}_\Sigma - \not{k} + m_\Sigma}{u - m_\Sigma^2}\gamma_5 \not{q}\right] u_p \ ,$$

where k and q are momentum of kaon and photon, ϵ is the photon polarization, F is the corresponding electromagnetic form factors (equals to 1 in the case of photoproduction), while μ_p, μ_Λ and μ_T represent the magnetic moments of proton, Λ-hyperon, and $\Sigma^0 - \Lambda$ transition [4]. For the K^+ and K_1 t-channel vector mesons the amplitudes can be written as

$$M_{\rm fi} = \bar{u}(p_\Lambda)\sum_{j=1}^{6} A_j(s, t, k^2)\ M_j\ u(p_p) \tag{2}$$

where s and t are the Mandelstam variables, while M_j are the gauge and Lorentz invariant matrices given by

$$\begin{aligned}
M_1 &= \frac{1}{2}\gamma_5(\not{\epsilon}\not{k} - \not{k}\not{\epsilon}) \\
M_2 &= \gamma_5[(2q - k)\cdot\epsilon P\cdot k - (2q - k)\cdot k\ P\cdot\epsilon] \\
M_3 &= \gamma_5(q\cdot k\not{\epsilon} - q\cdot\epsilon\not{k}) \\
M_4 &= i\epsilon_{\mu\nu\rho\sigma}\gamma^\mu q^\nu \epsilon^\rho k^\sigma \\
M_5 &= \gamma_5[q\cdot\epsilon k^2 - q\cdot k\ k\cdot\epsilon] \\
M_6 &= \gamma_5(k\cdot\epsilon\not{k} - k^2\not{\epsilon})
\end{aligned} \tag{3}$$

with $P = (p_p + p_\Lambda)/2$ and $\epsilon_{\mu\nu\rho\delta}$ is Levi-Civita tensor.

The last diagram shown in Fig. 1 is the contact term required in the pseudo-vector theory to maintain the gauge invariance of the amplitude. In this work, we use the contact term based on the previous work [5],

$$M^c = \frac{F^c g_{K\Lambda N}}{k^2(m_p + m_\Lambda)}\bar{u}_\Lambda\left[-\gamma_5(k^2\not{\epsilon} - k.\epsilon\not{k})\right] \tag{4}$$

Equation (2) presents the amplitude of kaon photo- and electroproduction. In the case of photoproduction, i.e., $k^2 = 0$, the matrices M_5 and M_6 do not exist and the formulation is significantly simplified.

LET assumes that kaon has a small mass compared to other baryon masses. To obtain the cross section at threshold, we approximate the three-momentum of kaon to zero, i.e., $|q| \approx 0$. The expansion of amplitude will be performed by using this approximation.

At $q \approx 0$, by using Eqs. (1), (2) and (3) it is obvious that $A_1(kq = m_k^2 = 0) = 0$. Therefore, we have to expand the A_1 in terms of m_k up to 2nd order. The function A_i can obtained from the previous study [4].

$$
\begin{aligned}
A_1 &= -\frac{G^T_{K*} t F^{K*}}{M(t - m^2_{K*} + i m_{K*}\Gamma_{K*})(m_p + m_\Lambda)} \\
A_2 &= \frac{G^T_{K*} F^{K*}}{M(t - m^2_{K*} + i m_{K*}\Gamma_{K*})(m_p + m_\Lambda)} - \frac{G^T_{K1} F^{K1}}{M(t - m^2_{K1} + i m_{K*}\Gamma_{K1})(m_p + m_\Lambda)} \\
A_3 &= -\frac{G^V_{K1} F^{K1}}{M(t - m^2_{K1} + i m_{K1}\Gamma_{K1})} \\
A_4 &= \frac{G^V_{K*} F^{K*}}{M(t - m^2_{K*} + i m_{K*}\Gamma_{K*})}
\end{aligned}
\tag{5}
$$

$M = 1$GeV is introduced to make coupling constants of K_1 and K^* dimensionless. Γ_i is width of each particle. The imaginary part of A_i will be ignored to simplify the calculation.

3. GENERAL RESULTS AND DISCUSSION

The analytic form of cross-section is calculated from the expansion of Eqs. (1), (2), (3) and (4) in terms of the ratio of kaon mass (493.677 MeV) and the average of nucleon (938.272 MeV) and Λ (1115.683 MeV) masses [7] ($m \approx m_p \approx m_\Lambda \approx m_\Sigma$).

$$
\begin{aligned}
|\mathcal{M}_{fi}|^2 = & -\Big[-6\, g^2_{K\Lambda N} + (6\, e\, g^2_{K\Lambda N} + 4\, A_1\, g_{K\Lambda N}\, m^2)\, x + (-\, g^2_{K\Lambda N} - 2e\, g^2_{K\Lambda N} \\
& -e^2\, g^2_{K\Lambda N} - 4\, g^2_{K\Lambda N}\, \mu_\Lambda\, m - 12e\, g^2_{K\Lambda N}\, \mu_P\, m - 4\, g_{K\Sigma N}\, g_{K\Lambda N}\, \mu_T\, m \\
& -8A_1 e\, g_{K\Lambda N} m^2 + 24 A_3\, g_{K\Lambda N} m^3 - 16 A_1^2 m^4)\, x^2 + (\, g^2_{K\Lambda N} + 2e\, g^2_{K\Lambda N} \\
& +e^2\, g^2_{K\Lambda N} + 128 A_4\, g_{K\Lambda N}\, m + 2\, g^2_{K\Lambda N}\, \mu_\Lambda\, m + 4e\, g^2_{K\Lambda N}\, \mu_\Lambda\, m \\
& +6e\, g^2_{K\Lambda N}\, \mu_P\, m + 4e^2\, g^2_{K\Lambda N}\, \mu_P\, m + 2\, g_{K\Lambda N}\, g_{K\Sigma N}\, \mu_T\, m \\
& +4e\, g_{K\Lambda N}\, g_{K\Sigma N}\, \mu_T\, m + 6 A_1\, g_{K\Lambda N} m^2 + 4 A_1 e\, g_{K\Lambda N} m^2 - 8 A_3\, g_{K\Lambda N} m^3 \\
& -8 A_3 e\, g_{K\Lambda N} m^3 + 16 A_1\, g_{K\Lambda N}\, \mu_\Lambda\, m^3 + 16 A_1 e\, g_{K\Lambda N}\, \mu_P\, m^3 \\
& +16 A_1\, g_{K\Sigma N}\, \mu_T\, m^3 - 32 A_1 A_3 m^5)\, x^3 + O(x^4)\Big]\ ,
\end{aligned}
\tag{6}
$$

where $x = m_K/m$, with m the average mass of nucleon and Λ, and $O(x^4)$ indicates the neglected higher orders.

The threshold cross-section of kaon photoproduction $\gamma p \longrightarrow K^+\Lambda$ is given by

$$
\frac{d\sigma}{d\Omega} = \frac{1}{64\pi^2\, s} \frac{|\mathbf{k}|}{|\mathbf{q}|} |\mathcal{M}_{fi}|^2 \tag{7}
$$

From the above result it is obvious that the contribution of A_4 is very small. Since the cross-section does not depend on A_2, only the A_1 and A_3 appear in the formula.

4. SUMMARY AND CONCLUSIONS

We have succesfully expanded the cross-section of kaon photoproduction at the threshold region up to the 3th order. As the follow up of this work, we will calculate the cross-section up to 5th order or more to refine the calculation. We will also try to compare the cross-section with experimental data near the threshold by using the particles properties listed in the particle listings of Particle Data Group (PDG) [7].

5. ACKNOWLEDGMENTS
This work has been partly supported by the PITTA Grant Universitas Indonesia No. 2320/UN2.R3.1/HKP.05.00/2018.

References

[1] Vainshtein A I and Zakharov V I 1972 *Nucl. Phys. B* **36** 589
[2] Gaffney G W 1967 *Phys. Rev.* **161** 1599
[3] Kroll N M and Ruderman M A 1954 *Phys. Rev.* **93** 233
[4] Mart T 2010 *Phys. Rev. C* **82** 025209
[5] Deo,B B and Bisoi, A K *Phys. Rev. D* **9** 288(1974)
[6] V.Koch. nucl-th/9512029
[7] Olive K A *et al.* 2014 *Chin. Phys.* C **38** 090001

Journal of Physics: Conference Series **1354** (2019) 012017 doi:10.1088/1742-6596/1354/1/012017

Theoretical Study of Unconventional Plasmon Generation by Solving Finite-Size 3D Hubbard Model within Mean Field Approximation

Dedi Prakasa, Muhammad Aziz Majidi[†]

[†]Department of Physics, Faculty of Mathematics and Natural Sciences, Universitas Indonesia, Kampus UI Depok, Depok 16424, Indonesia

Corresponding author's email: aziz.majidi@sci.ui.ac.id

Abstract. Recent experimental study on Strontium Niobate Oxide system has revealed unconventional plasmons generated due to confinement by oxygen planes. A phenomenological model accompanied the experimental data on that report has suggested that the confined electrons behave as harmonic oscillators. These motivate us to further study the effect of space confinement to the electrons on the formation of unconventional plasmons theoretically. For this purpose, we propose a method of modeling the generation of the unconventional plasmons in a finite system. The dynamics of correlated electrons in this confined system is described using finite-size 3D Hubbard model that is solved within mean field approximation. We study a hypothetical cubic system that consists of 5x5x5 single-orbital atoms. The calculation is done with and without incorporating the on-site Coulomb repulsion. We also consider both restricted and unrestricted mean field treatments to the calculation. Our dielectric function results show that for restricted mean field with U = 0 eV and unrestricted mean field with U $\leq$ 2 eV, the system is metal and only two and four unconventional plasmons found respectively. While, for U $>$ 2 eV, it becomes insulator and shows more emergence of unconventional plasmons.

1. Introduction

Plasmonics is a research field that studies the interaction between electromagnetic field and free electrons in metals. This field of research is promising due to the fact that subwavelength light confinement in metals lead to creation of novel devices [1]. The term plasmon refers to the collective oscillation of plasma, which is an electrically neutral medium with at least one type of charge is mobile [2]. According to Drude model, the plasma frequency can be defined as

$$\omega_p^2 = \frac{4\pi n e^2}{m\varepsilon_{core}} \tag{1}$$

where n is the total carrier density, m is the mass of electron and ε_{core} is the core dielectric constant.

A recent experimental study on strontium niobate oxide by Asmara *et al.* [3] shows that there exists plasmons which have different characteristics of real part $\varepsilon_1(\omega)$, imaginary part of complex dielectric function $\varepsilon_2(\omega)$, and loss function. We call these plasmons as unconventional plasmons which are generated due to oxygen planes confinement. Theoretical study of this

Content from this work may be used under the terms of the Creative Commons Attribution 3.0 licence. Any further distribution of this work must maintain attribution to the author(s) and the title of the work, journal citation and DOI.
Published under licence by IOP Publishing Ltd

confined system suggests that the electrons behave as harmonic oscillator. The results of that research motivates us to study the effect of confinement and electron-electron interaction to the generation of the unconventional plasmons. We develop a model of a hypothetical system that consists of 5x5x5 single-orbital atoms. Our goal of this research is to calculate the density of states, complex dielectric function, and loss function of the system.

2. Model

We demonstrate a system of 5x5x5 single-orbital atoms where the mobility of the electrons is limited due to the finite amount of sites.

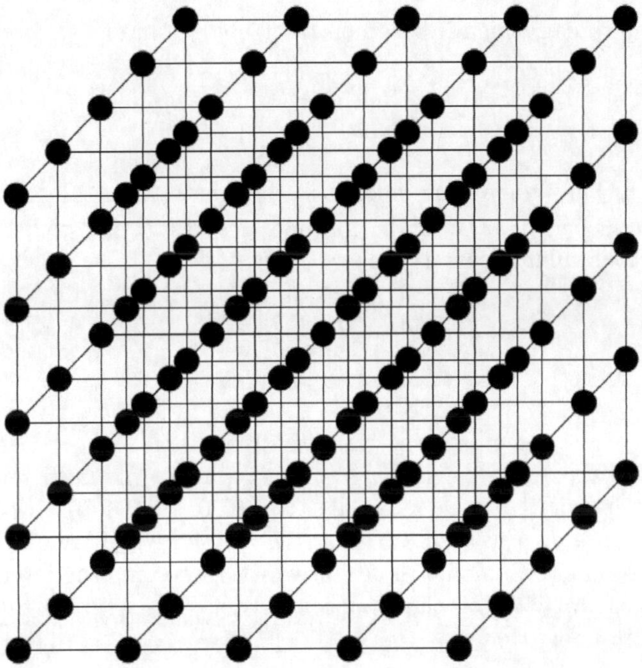

Figure 1. Illustration of the system where each dot represents a single-orbital atom.

Each site has index corresponds to the coordinate of the the site. Following are the list of indices and their corresponding coordinate (x,y,z)

$(0,0,0) \rightarrow 1$	$(1,0,0) \rightarrow 26$	$(2,0,0) \rightarrow 51$	$(3,0,0) \rightarrow 76$	$(4,0,0) \rightarrow 101$
$(0,0,1) \rightarrow 2$	$(1,0,1) \rightarrow 27$	$(2,0,1) \rightarrow 52$	$(3,0,1) \rightarrow 77$	$(4,0,1) \rightarrow 102$
$(0,0,2) \rightarrow 3$	$(1,0,2) \rightarrow 28$	$(2,0,2) \rightarrow 53$	$(3,0,2) \rightarrow 78$	$(4,0,2) \rightarrow 103$
$(0,0,3) \rightarrow 4$	$(1,0,3) \rightarrow 29$	$(2,0,3) \rightarrow 54$	$(3,0,3) \rightarrow 79$	$(4,0,3) \rightarrow 104$
$(0,0,4) \rightarrow 5$	$(1,0,4) \rightarrow 30$	$(2,0,4) \rightarrow 55$	$(3,0,4) \rightarrow 80$	$(4,0,4) \rightarrow 105$
$(0,1,0) \rightarrow 6$	$(1,1,0) \rightarrow 31$	$(2,1,0) \rightarrow 56$	$(3,1,0) \rightarrow 81$	$(4,1,0) \rightarrow 106$
$(0,1,1) \rightarrow 7$	$(1,1,1) \rightarrow 32$	$(2,1,1) \rightarrow 57$	$(3,1,1) \rightarrow 82$	$(4,1,1) \rightarrow 107$
$(0,1,2) \rightarrow 8$	$(1,1,2) \rightarrow 33$	$(2,1,2) \rightarrow 58$	$(3,1,2) \rightarrow 83$	$(4,1,2) \rightarrow 108$
$(0,1,3) \rightarrow 9$	$(1,1,3) \rightarrow 34$	$(2,1,3) \rightarrow 59$	$(3,1,3) \rightarrow 84$	$(4,1,3) \rightarrow 109$
$(0,1,4) \rightarrow 10$	$(1,1,4) \rightarrow 35$	$(2,1,4) \rightarrow 60$	$(3,1,4) \rightarrow 85$	$(4,1,4) \rightarrow 110$
$(0,2,0) \rightarrow 11$	$(1,2,0) \rightarrow 36$	$(2,2,0) \rightarrow 61$	$(3,2,0) \rightarrow 86$	$(4,2,0) \rightarrow 111$
$(0,2,1) \rightarrow 12$	$(1,2,1) \rightarrow 37$	$(2,2,1) \rightarrow 62$	$(3,2,1) \rightarrow 87$	$(4,2,1) \rightarrow 112$
$(0,2,2) \rightarrow 13$	$(1,2,2) \rightarrow 38$	$(2,2,2) \rightarrow 63$	$(3,2,2) \rightarrow 88$	$(4,2,2) \rightarrow 113$
$(0,2,3) \rightarrow 14$	$(1,2,3) \rightarrow 39$	$(2,2,3) \rightarrow 64$	$(3,2,3) \rightarrow 89$	$(4,2,3) \rightarrow 114$

$(0,3,0)\to 16$	$(1,3,0)\to 41$	$(2,3,0)\to 66$	$(3,3,0)\to 91$	$(4,3,0)\to 116$
$(0,3,1)\to 17$	$(1,3,1)\to 42$	$(2,3,1)\to 67$	$(3,3,1)\to 92$	$(4,3,1)\to 117$
$(0,3,2)\to 18$	$(1,3,2)\to 43$	$(2,3,2)\to 68$	$(3,3,2)\to 93$	$(4,3,2)\to 118$
$(0,3,3)\to 19$	$(1,3,3)\to 44$	$(2,3,3)\to 69$	$(3,3,3)\to 94$	$(4,3,3)\to 119$
$(0,3,4)\to 20$	$(1,3,4)\to 45$	$(2,3,4)\to 70$	$(3,3,4)\to 95$	$(4,3,4)\to 120$
$(0,4,0)\to 21$	$(1,4,0)\to 46$	$(2,4,0)\to 71$	$(3,4,0)\to 96$	$(4,4,0)\to 121$
$(0,4,1)\to 22$	$(1,4,1)\to 47$	$(2,4,1)\to 72$	$(3,4,1)\to 97$	$(4,4,1)\to 122$
$(0,4,2)\to 23$	$(1,4,2)\to 48$	$(2,4,2)\to 73$	$(3,4,2)\to 98$	$(4,4,2)\to 123$
$(0,4,3)\to 24$	$(1,4,3)\to 49$	$(2,4,3)\to 74$	$(3,4,3)\to 99$	$(4,4,3)\to 124$
$(0,4,4)\to 25$	$(1,4,4)\to 50$	$(2,4,4)\to 75$	$(3,4,4)\to 100$	$(4,4,4)\to 125$

We propose a hamiltonian which is based on tight binding model as follows

$$H = \sum_i a^\dagger_{i,\sigma}[H_0]a_{j,\sigma} + U\sum_i n_{i\uparrow}n_{i\downarrow} \tag{2}$$

The first term of the hamiltonian corresponds to the kinetic term. $a^\dagger_{i,\sigma}$ and $a_{j,\sigma}$ are row vector of creation and coloumn vector of annihiliation operator respectively. $H[0]$ is a 250x250 that can be grouped into four submatrices as follows

$$[H_0] = \begin{bmatrix} [H_0]_\uparrow & \mathbf{0} \\ \mathbf{0} & [H_0]_\downarrow \end{bmatrix} \tag{3}$$

where $[H_0]_{\uparrow(\downarrow)}$ diagonal elements are the on-site energy ε. Other elements that have index representing two neighbouring site where the difference of the coordinates $(|x - x'|, |y - y'|, |z - z'|)$ equals to $(0,0,1), (0,1,0)$, or $(1,0,0)$ such as $H_{1,2}, H_{1,6}, H_{1,26}, H_{2,1}, H_{2,3}, H_{2,7}, H_{2,27}, ..., H_{125,100}, H_{125,120}, H_{125,124}$ have value of $-t$.

The second term is on-site Coulomb repulsion which represents the interaction between two electrons in the same orbital with U is the repulsion energy. This term is treated using mean-field approximation and can be written as

$$U\sum_i n_{i\uparrow}n_{i\downarrow} = U\sum_i \Big(\langle n_{i\uparrow}\rangle n_{i\downarrow} + \langle n_{i\downarrow}\rangle n_{i\uparrow} - \langle n_{i\uparrow}\rangle\langle n_{i\downarrow}\rangle\Big) \tag{4}$$

where the occupation number $\langle n_{i\uparrow}\rangle$ and $\langle n_{i\downarrow}\rangle$ are calculated using self-consistent method.

3. Methods of Calculation

We define Green's function to calculate the density of states of the system. In matrix form, the Green's function can be written as

$$[G(z)] = \Big[z[I] - [H_0] - [\Sigma(z)]\Big]^{-1} \tag{5}$$

where z is complex frequency variable defined as $\omega + i0^+$ and $[I]$ is a 125×125 identity matrix. The self-energy matrix, $[\Sigma(\omega)]$ contains interaction terms of the Hamiltonian and it will be solved using self-consistent method. Elements of $[\Sigma(\omega)]$ are all zero except for the diagonal elements Σ_i which has value of

$$\Sigma_1 = \Sigma_2 = \Sigma_3 = \ldots = \Sigma_{125} = U\langle n_{i\downarrow}\rangle \tag{6}$$

$$\Sigma_{126} = \Sigma_{127} = \Sigma_{128} = \ldots = \Sigma_{250} = U\langle n_{i\uparrow}\rangle \tag{7}$$

where the value of $\langle n_{i\uparrow(\downarrow)}\rangle$ is determined by certain initially guessed value.

From the Green's function, we can calculate the density of states (DOS) of the system through

$$DOS(\omega) = -\frac{1}{\pi} ImTr[G(\omega + i0^+)] \tag{8}$$

Average electron occupancy at site i with spin σ can be calculated through

$$\langle n_{i\sigma}\rangle = \int d\omega PDOS_{i\sigma}(\omega) f(\omega, \mu, T) \tag{9}$$

where $PDOS_{i\sigma}$ is defined as

$$PDOS_{i\sigma} = -\frac{1}{\pi} Im[G_{i\sigma}(\omega + i0^+)] \tag{10}$$

The calculated $\langle n_{i\sigma}\rangle$ are then used to be the new self-energy matrix elements. This updated self-energy matrix is fed back into Green's function as we defined in equation (5). We iterate this process until convergenced is achieved.

We calculate the optical conductivity for T = 0 K through

$$Re\sigma_{\alpha\beta} = \frac{\pi e^2}{\hbar V} \sum_k Tr[\nu_{\alpha\beta}][A_{\alpha\beta}(\mu)][\nu_{\alpha\beta}][A_{\alpha\beta}(\mu)] \tag{11}$$

Real and imaginary part of complex dielectric function $\varepsilon_1(\omega)$ and $\varepsilon_2(\omega)$ can be calculated using Kramers-Kronig formula

$$\varepsilon_1(\omega) - 1 = \frac{2}{\pi} \int_0^\infty \frac{\omega' \varepsilon_2(\omega')}{\omega'^2 - \omega^2} d\omega' \tag{12}$$

$$\varepsilon_2(\omega) = -\frac{2}{\pi} \int_0^\infty \frac{\omega' \varepsilon_2(\omega')}{\omega'^2 - \omega^2} d\omega' \tag{13}$$

From the ε_1 and ε_2, we can define loss function as.

$$-Im[\varepsilon^{-1}(\omega)] = \frac{\varepsilon_2}{\varepsilon_1^2 + \varepsilon_2^2} \tag{14}$$

4. Results and Discussion

We set the value of the parameters as follows: $\varepsilon = 0$ eV, $t = 1$ eV, $T = 0$ K, and vary the value of $U = 0$ eV, 0.5 eV, 1 eV, 2 eV, 3 eV, and 4 eV. Mean field approximation where $U = 0$ eV is said to be unrestricted, while for $U > 0$ eV is restricted due to the initial guess given for the occupation number. We work with half-filled electron system by setting $n_{filling}$ to be 125 electrons.

By ignoring the on-site Coulomb interaction between electrons, we aim to see only the effect of confinement to the formation of unconventional plasmons. We can see from dielectric function curve that it has the characteristic of multi-oscillator model [4] and interband transition due to bound charges [5, 6, 7] as at 0.75 and 1 eV. Intense peak arises from loss function curve shows the characteristic of conventional plasmons due to its correspondence to the point where $\varepsilon_1 = 0$. Other conventional plasmons can also be found at low frequency (0.1 eV and 0.8 eV) as shown in Figure 3. Setting U = 0.5 and 1 eV, resulting in lowering the magnitude of the peaks and slightly shifting the loss function. Therefore, conventional plasmons are no longer found at low frequency.

Journal of Physics: Conference Series **1354** (2019) 012017 doi:10.1088/1742-6596/1354/1/012017

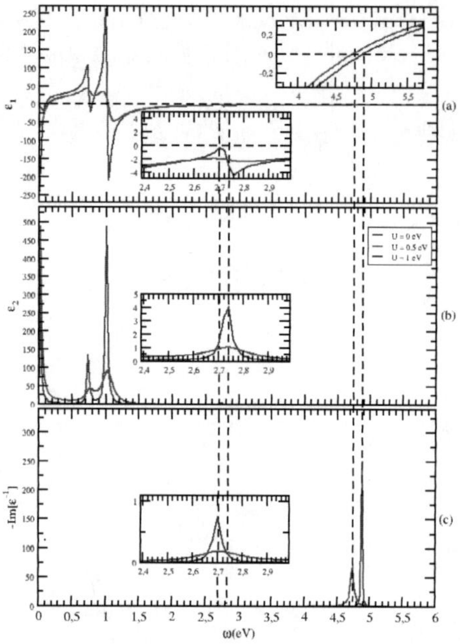

Figure 2. Complex dielectric function and loss function for U = 0 eV, 0.5 eV, and 1 eV. Upper subplot of subfigure 2(a) shows the shift of conventional plasmon. While middle subplot of figure 2(a), 2(b), and 2(c) shows the peaks of conventional and unconventional plasmons.

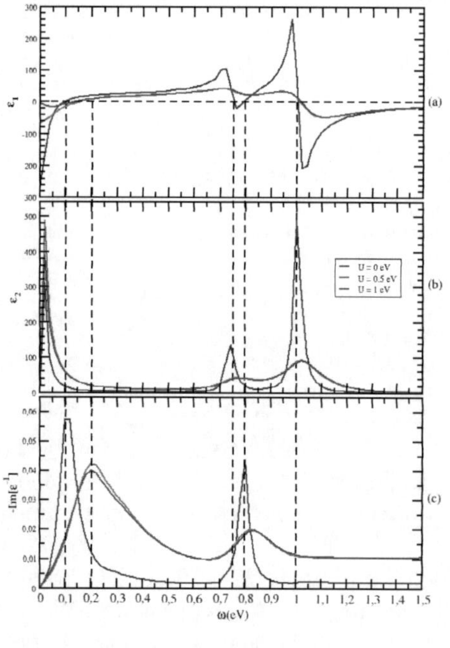

Figure 3. A more careful inspection of Figure 2 on lower frequency shows the emergence of both conventional and unconventional plasmons.

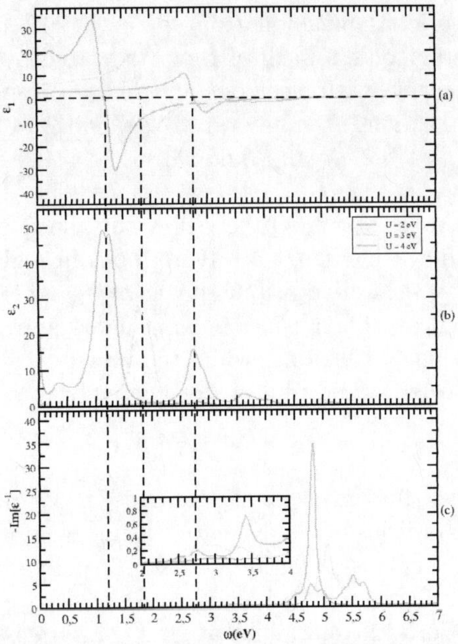

Figure 4. Real and imaginary part of complex dielectric function and loss function for U = 2, 3, and 4 eV. Unconventional plasmons can bee seen from peaks at subfigure 4(b) which aligned with dashed line.

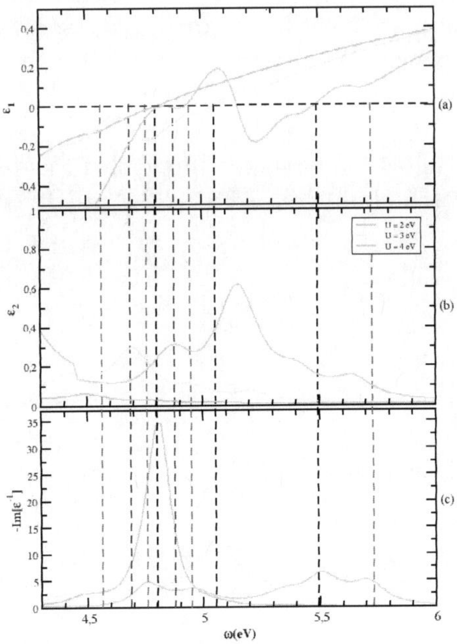

Figure 5. More conventional and unconventional shown in region between 4.5 eV to 6 eV. Black dashed lines indicate conventional plasmons, while red and blue for unconventional plasmons which weakly interact with electrons and photons respectively.

Here we identify unconventional plasmon from ε_2 or loss function peak which does not correspond to zero crossing point of ε_1. Each of them indicated by ε_2 and loss function peaks, is said to be inactive to interaction with electrons and photon respectively. From Figure 2, we can see that unconventional plasmons are found at 2.7 eV and 2.74 eV for U = 0 eV. Two more unconventional plasmons arise at 0.2 eV and 0.85 eV for U = 0.5 eV and 1 eV as can be seen from Figure 3.

For U = 2, 3, and 4 eV, the ε_1 and ε_2 curves show the characteristic of complex dielectric function for insulator [6] as shown in Figure 4 . If we focus in region between 4.5 - 6 eV as in Figure 5, we see that there exist several additional peaks of ε_2 and loss function curve, compared to Figure 2. Those peaks indicate unconventional plasmons as we defined previously. Black dashed lines indicate conventional plasmons, while red and blue for unconventional plasmons which weakly interact with electrons and photons respectively.

5. Conclusion
In conclusion, the emergence of unconventional plasmons in our model using Hubbard model for system of 5x5x5 single-orbital atoms and treated using mean-field approximation, can be seen from peaks arise from ε_2 and loss function curve. We define two types of unconventional plasmons, the first one is identified by peak from ε_2 and the other one from loss function curve which does not correspond to zero-crossing point of ε_1. Each type is inactive to interaction with electrons and photons respectively.

Calculation of unrestricted mean field in our model shows characteristic of metal and Unconventional plasmons are found at 2.7 and 2.74 eV. While, for restricted mean field with U = 0.5 eV and 1 eV, results additional unconventional plasmons at 0.2 eV and 0.85 eV. For higher Coulomb repulsion, U = 2 eV, 3 eV, and 4 eV, the system turns into insulator and several unconventional plasmons are identified in frequency range of 4.5 eV to 6 eV. Therefore, confinement of electrons in our model is able to generate unconventional plasmons.

6. Acknowledgement
We would like to thank Universitas Indonesia for supporting this research by fully funding this project through PITTA Research Grant No. 2285/UN2.R3.1/HKP.05.00/2018

References
[1] West, P.R., Ishii, S., Naik, G.V., Emani, N.K., Shalaev, V.M., and Boltasseva, A. (2010) Searching for better plasmonic materials. Laser Photonics Rev., 4, 795808
[2] Maier, S.A. 2007 Plasmonics: Fundamentals and Applications. Springer
[3] Asmara T.C. et al. 2017 Tunable and low-loss correlated plasmons in Mott-like insulating oxides. Nat. Commun. 8, 15271
[4] Novotny L. and Hecht B. 2006 Principles of Nano-Optics. Cambridge University Press, pp. 381-382
[5] Stenzel O. 2016 The Physics of Thin Film Optical Spectra: An Introduction. Springer, pp. 35-39
[6] Vos M, and Grande P.L. 2016 Simple model dielectric functions for insulators. Journal of Physical and Chemistry of Solids, 104 pp. 192-197
[7] Wysin G.M., Chikan V., Young N., Dani R.K. 2013 Effects of interband transitions on Faraday rotation in metallic nanoparticles. J Phys. Condens. Matter 25:325302

Drop of Spin Polarization of Half Metal due to Interfacial Spin Mixing

Adam B. Cahaya and M. A. Majidi
Department of Physics, Faculty of Mathematics and Natural Sciences Universitas Indonesia, Kampus UI Depok, Depok 16424, Indonesia
E-mail: aziz.majidi@sci.ui.ac.id

Abstract. Half metal is a ferromagnetic metal that act as a conductor to electrons with of one spin orientation, while act as an insulator for the opposite spin. Because of this high spin-dependent property, half metals are widely studied in spintronics area. This property can be represented by the spin polarization $P = (G_\uparrow - G_\downarrow)/(G_\uparrow + G_\downarrow)$ of its spin-dependent conductances $G_{\uparrow,\downarrow}$. The value of P of an ideal half metal is 1. However, the value of P is reduced when it is attached to other material. We study how the spin current generations that occurs at the interface of half metal and normal metal reduces the value P. At the interface of a half metal and a normal metal a spin current can be pumped from the ferromagnetic layer to the normal metal. This spin pumping effect is characterized by spin-mixing conductance, which represents the magnitude of the generated spin current due to the mixing of electrons with spin up and down. By analyzing the coupling of spin and electric currents in the interface of half metal and normal metal, we show that the spin mixing strongly influences the spin-dependent transport near the interface, and therefore reduces the spin polarization of the half metal.

1. Introduction

Half metal is a ferromagnet with a strong spin-dependent properties[4]. At the Fermi level, the density of state of one of its spin is equal to zero. As a consequent, half-metallic material acts as a conductor only to electrons with one spin orientation. In a two-current model, this property is measured by a spin polarization

$$P = \frac{G_\uparrow - G_\downarrow}{G_\uparrow + G_\downarrow}, \tag{1}$$

where $G_\uparrow$ and $G_\downarrow$ are the conductances of the majority and minority spin, respectively. For an ideal half metal $G_\uparrow \gg G_\downarrow$, therefore $P = 1$.

Spin-dependent band calculations have shown that Heusler alloys and half-Heusler alloys have half-metallic band structures, *i.e.* the density of states of one spin is zero at Fermi level [7, 6]. However, the measured values of spin polarization in experiments has shown values $P < 1$. The measurements of less than 1 values of P may come from the influences of adjacent materials attached to the half metal. Among the phenomena that occurs at the interface of a half metal and an adjacent non-magnetic metal, a recent experiment showed that a pure spin current, a pure angular momentum flow without any electric currents, is transferred through the interface [1]. The quantity that directly related to the spin current generation in Ref. [1] is the spin mixing conductance $g_{\uparrow\downarrow}$. The spin-mixing conductance is the property of the interface where the spin current is generated [13]. While the spin current can also be generated by ferromagnetic

Content from this work may be used under the terms of the Creative Commons Attribution 3.0 licence. Any further distribution of this work must maintain attribution to the author(s) and the title of the work, journal citation and DOI.
Published under licence by IOP Publishing Ltd

resonance [10] or electric voltage [9, 16], the generated spin currents are governed by the same spin-mixing conductances.

Unlike electric current, which is a vector, a spin current is a tensor that can be characterized with a current and a polarization vectors that represents the direction of its movement and its angular momentum direction, respectively [12]. In this article, we use the following definition for spin current tensor

$$\mathbb{J}_s = \frac{1}{2}\left(\mathbf{j}_s \otimes \boldsymbol{\sigma} + \boldsymbol{\sigma} \otimes \mathbf{j}_s\right), \tag{2}$$

where $\boldsymbol{\sigma}$ is the polarization vector and $\mathbf{j}_s$ is the spin current vector.

In this article, we focuses on the role of the spin-mixing conductance in reducing the measured value of spin polarization P. By utilizing the two-currents model of majority and minority spins [5, 11], we analyze the coupling of the spin and electric currents in the bilayer of a half metal and a non-magnetic metal. The pure spin current generation at the interface reduces the coupling of the spin and electric currents, thus, reduces the measured spin polarization.

2. Spin current distribution

To determine the spin polarization of the conductance in a real system that includes interfaces, we need to consider a ferromagnet (F) sandwiched by non-magnetic metals (N) at $x = \pm l/2$ under an applied voltage V. Here l is the width of the ferromagnet. The non-magnetic metals are assumed to have a large conductance, compared to F, such that the voltage drop in N can be ignored.

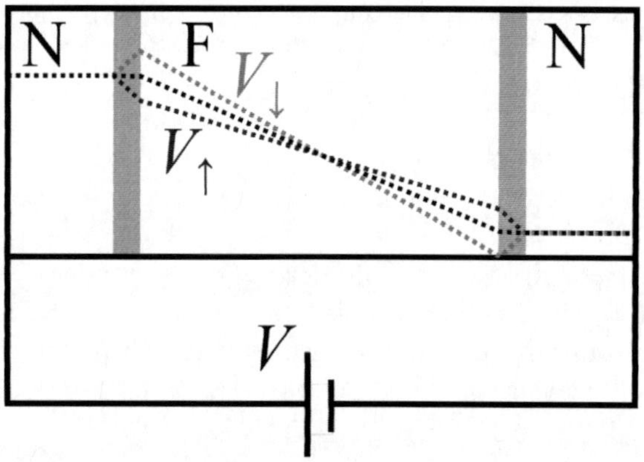

Figure 1. To see the effect of the interface, we consider a ferromagnet (F) sandwiched by non-magnetic metals (N). The large conductances of the non-magnetic metals induces negligible voltage drop at both Ns. In F, the voltage become spin-dependent $V_\uparrow$ and $V_\downarrow$, which corresponds to the majority and minority spins respectively.

2.1. Coupling of spin and charge current in F

We start from a more general ferromagnet with spin-polarization $P_F \leq 1$, the spin-dependent transport can be represented in the following current-force matrix relation

$$\begin{pmatrix} \mathbf{j} \\ \mathbf{j}_s \end{pmatrix} = \sigma_F \begin{pmatrix} 1 & P_F \\ P_F & 1 \end{pmatrix} \begin{pmatrix} -\nabla V \\ -\nabla\left(\boldsymbol{\sigma} \cdot \mathbf{V}_s/2\right) \end{pmatrix}, \tag{3}$$

Journal of Physics: Conference Series **1354** (2019) 012018 doi:10.1088/1742-6596/1354/1/012018

where σ_F is the conductivity of F, $V = (V_\uparrow + V_\downarrow)/2$ and $V_s = V_\uparrow - V_\downarrow$ is the electric voltage and spin accumulation, respectively, which are related to the spin dependent voltages. $V_\uparrow$ and $V_\downarrow$ are the voltage of the spin up $\mathbf{j}_\uparrow$ and spin down $\mathbf{j}_\downarrow$ current densities, respectively. Here we limit our discussion of coupling of charge $\mathbf{j} = \mathbf{j}_\uparrow + \mathbf{j}_\downarrow$ and spin $\mathbf{j}_s = \mathbf{j}_\uparrow - \mathbf{j}_\downarrow$ densities in the absence of thermal gradient [2, 8].

Due to spin-flip scattering, the spin current is diffused through F. The diffusion relation can be written in term of a diffusive spin accumulation

$$\nabla^2 \mathbf{V}_s = \frac{\mathbf{V}_s}{\lambda^2}, \tag{4}$$

where λ is the spin-diffusion length. Substituting Eq. 3, the solution is

$$\mathbf{V}_s = \mathbf{V}_0 \frac{\sinh \frac{x}{\lambda}}{\sinh \frac{l}{2\lambda}}. \tag{5}$$

Here we applied the boundary condition of $\mathbf{V}_s(\pm l/2) = \pm \mathbf{V}_0$. Setting $V(0) = 0$, the electric voltage distribution can also be obtained

$$V(x) = -\frac{|\mathbf{j}|\, x}{\sigma} - P_F \boldsymbol{\sigma} \cdot \mathbf{V}_0 \frac{\sinh \frac{x}{\lambda}}{\sinh \frac{l}{2\lambda}} \tag{6}$$

2.2. Spin current generation in F | N interface

At the interface, spin currents also arised due to spin mixing conductance $g^{\uparrow\downarrow} = g_r + i g_i$, namely spin currents associated with spin transfer torques J_{ST} and spin pumping effects J_{SP}. J_{SP} arises from the dynamics of the magnetization vector $\mathbf{m}$ [13]

$$j_{\mathrm{SP}} \boldsymbol{\sigma} \propto g_r \mathbf{m} \times \dot{\mathbf{m}} - g_i \dot{\mathbf{m}}. \tag{7}$$

Proportionality indicates a conversion constant from spin current in Joule unit to Ampere unit. On the other hand, spin transfer torque arises from the relative orientation of directions of magnetization and of spin accumulation [3, 8].

$$j_{\mathrm{ST}} \boldsymbol{\sigma} \propto g_r \mathbf{m} \times (\mathbf{m} \times \Delta \mathbf{V}_s) - g_i \mathbf{m} \times \Delta \mathbf{V}_s, \tag{8}$$

where $\Delta \mathbf{V}_s = \mathbf{V}_s^N - \mathbf{V}_s^F$ is the spin accumulation drop at the interface.

Since generally $\mathrm{Re} g^{\uparrow\downarrow} \gg \mathrm{Im} g^{\uparrow\downarrow}$, we will focus on the g_r term. In steady state, j_{SP} is balanced by backflow of spin current due to (Johnson-Nyquist) thermal fluctuation [15]. Because of that, the interface spin generation effects contribution to the spin current is

$$j_s^{\uparrow\downarrow} \boldsymbol{\sigma} = -g_s \Delta \mathbf{V}_s, \tag{9}$$

where g_s is proportional to g_r [14]

$$g_s = g_r \frac{e \gamma \hbar Z_{3/2}}{\pi M_s d \lambda^3}, \tag{10}$$

γ is the gyromagnetic ratios, M_s saturated magnetization, d is thickness of the ferromagnet, $\lambda \propto T^{-1/2}$ is the de Broglie thermal wavelength of magnon and Z_n is the Zeta function.

When the spin accumulation in the N side is zero (see Fig. 1), the spin current at the interface is then

$$j_s\left(\pm\frac{l}{2}\right) = \mp g_s \boldsymbol{\sigma} \cdot \mathbf{V}_s\left(\pm\frac{l}{2}\right) = g_s \boldsymbol{\sigma} \cdot \mathbf{V}_0. \tag{11}$$

3. Result and Discussion

From the boundary condition for spin current at $\mathbf{j}_s\,(\pm l/2)$, we can substitute Eq. 6 to Eq. 3 and get the spin and electric current that is measured by the N probes.

$$|\mathbf{j}_s| = |\mathbf{j}| \frac{P_F}{1 + \dfrac{1 - P_F^2}{g_s \sigma_F \lambda} \coth \dfrac{l}{2\lambda}}, \tag{12}$$

where

$$|\mathbf{j}| = \frac{V}{\dfrac{l}{\sigma_F} + \dfrac{2P_F}{g_s + \sigma_F \dfrac{1 - P^2}{\lambda} \coth \dfrac{l}{2\lambda}}}. \tag{13}$$

The effective spin polarization can be determine by from the spin-dependent conductances

$$G_\uparrow \equiv \frac{|j_\uparrow|\, A}{V_\uparrow} \tag{14}$$

$$G_\downarrow \equiv \frac{|j_\downarrow|\, A}{V_\downarrow}, \tag{15}$$

where A is the cross-sectional area. Since the voltage is no longer spin dependent at the non-magnetic probes (see Fig. 1), the effective polarization is then

$$P = \frac{|j_\uparrow| - |j_\downarrow|}{|j_\uparrow| + |j_\downarrow|} = \frac{|j_s|}{|j|} = \frac{P_F}{1 + \dfrac{g_F}{g_s} \coth \dfrac{l}{2\lambda}}, \tag{16}$$

where $g_F = \sigma_F \lambda/(1 - P_F^2)$ is the conductance of F that responds to the spin current. P is determined by g_F/g_s ratio and the length of F. $\coth l/(2\lambda)$ dependency indicates that larger l felt smaller the polarization drop *i.e.* smaller interface effect.

4. Conclusion

We discussed the current transports in multilayer system involving half metals and non-magnetic metals. By taking the limit of very thin non-magnetic metals, we showed that the value of the spin polarization drops because of the interfacial effects represented by G_S, which is proportional to spin mixing conductance.

When spin and electric current do not couple in the interface, the spin polarization will not decrease. We have shown that when the spin current can penetrate the ferromagnetic bulk and therefore couple to electric current, it can reduce the effective spin polarization of the system.

Acknowledgement

This work was supported by Indonesian Ministry of Research, Technology and Higher Education PDUPT Research Grant No. 374/UN2.R3.1/HKP05.00/2018

Reference

[1] S. Bosu, Y. Sakuraba, K. Uchida, K. Saito, T. Ota, E. Saitoh, and K. Takanashi. Spin seebeck effect in thin films of the heusler compound co$_2$mnsi. *Phys. Rev. B*, 83:224401, 2011.

[2] A.B. Cahaya, O.A. Tretiakov, and G.E.W. Bauer. Spin seebeck power conversion. *IEEE Trans. Magn.*, 51:0800414, 2015.

[3] Y.-T. Chen, S. Takahashi, H. Nakayama, M. Althammer, S.T.B. Goennenwein, E. Saitoh, and G.E.W. Bauer. Theory of spin hall magnetoresistance. *Phys. Rev. B*, 87:144411, 2013.

[4] R. A. de Groot, F. M. Mueller, P. G. van Engen, and K. H. J. Buschow. New class of materials: Half-metallic ferromagnets. *Phys. Rev. Lett.*, 50:2024–2027, Jun 1983.

[5] A. Fert, P. Grnberg, A. Barthlmy, F. Petroff, and W. Zinn. Layered magnetic structures: interlayer exchange coupling and giant magnetoresistance. *J Magn. Magn. Mater.*, 140-144:1 – 8, 1995. International Conference on Magnetism.

[6] I. Galanakis, P. H. Dederichs, and N. Papanikolaou. Slater-pauling behavior and origin of the half-metallicity of the full-heusler alloys. *Phys. Rev. B*, 66:174429, 2002.

[7] T. Graf, C. Felser, and S.S.P. Parkin. Simple rules for the understanding of heusler compounds. *Prog. Solid State Chem.*, 39(1):1, 2011.

[8] M. Hatami, G.E.W. Bauer, Q. Zhang, and P.J. Kelly. Thermal spin-transfer torque in magnetoelectronic devices. *Phys. Rev. Lett.*, 99:066603, 2007.

[9] Y Ji, A Hoffmann, J S Jiang, J E Pearson, and S D Bader. Non-local spin injection in lateral spin valves. *Journal of Physics D: Applied Physics*, 40(5):1280–1284, feb 2007.

[10] R.H. Silsbee, A. Janossy, and P. Monod. Coupling between ferromagnetic and conduction-spin-resonance modes at a ferromagnetic¯normal-metal interface. *Phys. Rev. B*, 19:4382–4399, 1979.

[11] J.C. Slonczewski. Current-driven excitation of magnetic multilayers. *J Magn. Magn. Mater.*, 159(1):L1, 1996.

[12] Qing-feng Sun, X. C. Xie, and Jian Wang. Persistent spin current in nanodevices and definition of the spin current. *Phys. Rev. B*, 77:035327, Jan 2008.

[13] Y. Tserkovnyak, A. Brataas, and G.E.W. Bauer. Enhanced gilbert damping in thin ferromagnetic films. *Phys. Rev. Lett.*, 88:117601, 2002.

[14] J. Xiao and G. E. W. Bauer. Transport between metals and magnetic insulators. arXiv:1508.02486, 2015.

[15] J. Xiao, G.E.W. Bauer, K. Uchida, E. Saitoh, and S. Maekawa. Theory of magnon-driven spin seebeck effect. *Phys. Rev. B*, 81:214418, 2010.

[16] T. Yang, T. Kimura, and Y. Otani. Giant spin-accumulation signal and pure spin-current-induced reversible magnetization switching. *Nature Physics*, 4:851 EP –, Oct 2008.

Journal of Physics: Conference Series **1354** (2019) 012019 doi:10.1088/1742-6596/1354/1/012019

Perturbative and Numerical Approach on The Quantum Harmonic Oscillator with a Negative Gaussian Potential

Tony Sumaryada[1]*, Firdha Najiya[1], Sidikrubadi Pramudito[1], Akhiruddin Maddu[1], Multazimul Haq[1], Sugianto Arjo[2], Mirzanur Hidayat[2]
[1]Department of Physics, Bogor Agricultural University, Jalan Meranti Kampus IPB Dramaga Bogor 16680, Indonesia
[2]Program Studi Pendidikan Fisika, Fakultas Keguruan dan Ilmu Pendidikan, Universitas HAMKA, Jl. Tanah Merdeka, Pasar Rebo, Jakarta Timur, Jakarta 13830, Indonesia

E-mail : tsumaryada@ipb.ac.id

Abstract. Modeling a system in a harmonic oscillator approach still beneficial for the researcher to elucidate the physical mechanism in a small system such as a quantum dot. In this paper, we investigated the quantum harmonic oscillator system with the addition of a negative Gaussian potential (QHONGP). The effect of this Gaussian term was calculated analytically using a perturbative approach, and numerically using the Numerov and shooting method. The results showed that both approaches were able to show the unique feature of a pairing formation, i.e the decreasing energy of the system and the localization of particles in all (odd and even) states, at the lower pairing strength. For the higher pairing strength, the perturbative results highly depend on the number of energy levels and the expansion terms included in the calculation. The limited number of energy levels and the expansion terms resulted in the inaccuracy of the perturbative approach in giving the solution of QHONGP system as compared to the shooting method.

Keywords: Gaussian potential, harmonic oscillator, Numerov and shooting method, perturbation, quantum dot

1. Introduction

Quantum harmonic oscillator is the most common approach used by researchers to study the properties of a physical system in nature. The simplicity of the system and the wide range of problems modeled by this approach has made this system as an almost-default system in approaching many problems in nature. It had been used to model diatomic molecules, DNA, nucleons, quarks, economics, sociology and many more [1–5].

Quantum harmonic oscillator has been coupled with a various type of potentials for various purposes. It has been coupled with a positive Gaussian potential (a double-well potential) [6] and had successfully described the NH_3 molecule degeneracy in molecular physics [7,8]. In condensed matter, the Gaussian potential had been used as a new confinement potential for spherical quantum dots [9] and also for electron pairs system [10].

Content from this work may be used under the terms of the Creative Commons Attribution 3.0 licence. Any further distribution of this work must maintain attribution to the author(s) and the title of the work, journal citation and DOI.
Published under licence by IOP Publishing Ltd

The pairing formation is a common phenomenon that occurred in nature as a consequence of many-body interactions. This phenomenon manifest itself in any scale of physics, from superconductivity of electrons in the solid state, Bose-Einstein condensate in dilute atomic gas, or even superfluidity in nuclei and compact stars. The common features of pairing are the decreasing of the total energy of the system and the localization of particles. Those properties usually studied and obtained by solving the pairing problem within the framework of BCS theory which deploys mechanical operators based on the second quantization. The pairing bound state formation alternatively can be solved by using a regular quantum mechanics procedure in the harmonic and anharmonic oscillator basis, as shown by some researchers [11–13].
In this paper, we will demonstrate the ability and limitation of the perturbative and numerical approaches in describing the pairing formation in a quantum harmonic oscillator with a negative Gaussian potential (QHONGP) model. The perturbative approach is limited only up to the first five terms of the expansion and the analysis was done for the ground state up to the fourth excited state. The analysis will be focused on the energetic aspect of pairing and the wavefunction's profile of each state. Some advantages and limitations of those approaches will also be discussed here.

2. Methods

We start solving the QHONGP by writing the Hamiltonian of the problem as follows

$$H = \frac{\hat{p}^2}{2\mu} + \frac{1}{2}\mu\omega^2 x^2 - g e^{-\frac{x^2}{\sigma^2}} \tag{1}$$

where $\hat{p} = -i\hbar\partial/\partial x$ is the momentum operator, and $\mu = m/2$ is the reduced mass of harmonic oscillator of two identical mass system. For simplicity, we set $\hbar = \omega = \mu = \sigma = 1$. The constant g is the pairing strength and denotes the strength of the external perturbing potential. The parameters ω, σ represent the angular speed and the deviation's width of Gaussian potential respectively.

The Hamiltonian of the system due to the introduction of small perturbation around the equilibrium point of the system can be expressed as $H = H_0 + \lambda H_1$ with H_0 the unperturbed Hamiltonian and λH_1 the perturbing potential term. The first order correction of energy was calculated using the equation below which can be found in many quantum mechanics textbook [14].

$$E_n^1 = \langle \psi_n^0 | H' | \psi_n^0 \rangle \tag{2}$$

and the corresponding first order correction on the wave function is

$$\psi_n^1 = \sum_{m \neq n} \frac{\langle \psi_m^0 | H' | \psi_n^0 \rangle}{(E_n^0 - E_m^1)} \psi_m^0 \tag{3}$$

Note that m and n denote the two interacting levels. The wave functions of harmonic oscillator are expressed in terms of *Hermite* polynomial H_k and take this form,

$$\psi_k = \sqrt{\frac{1}{\sqrt{\pi}2^k k!}} H_k e^{-\frac{x^2}{2}} \tag{4}$$

where H_k for several terms (k is the polynomial's order) are shown below

$$H_0(y) = 1\;;\; H_1(y) = 2y\;;\; H_2(y) = 4y^2 - 2\;;\; H_3(y) = 8y^3 - 12y\;;$$
$$H_4(y) = 16y^4 - 48y^2 + 12 \tag{5}$$

The expansion is limited only up to five terms, those five terms (k =0, 1, 2, 3, 4) were assumed to be adequate enough to represent the dominant terms in the perturbative approach of the system.
For the second order correction, the energy correction follows this form

$$E_n^2 = \sum_{m \neq n} \frac{\left|\left\langle \psi_m^0 \middle| H' \middle| \psi_n^0 \right\rangle\right|^2}{E_n^0 - E_m^2} \tag{6}$$

By solving Equation (6) above, the total energy of each state can be calculated by adding the first and second order correction to the unperturbed energy of each state.
Solving Equation (1) can also be done numerically using the Numerov method combined with shooting method [15]. One needs to rewrite the potential term as

$$V(x) = \frac{1}{2}\mu\omega^2 x^2 - g e^{-\frac{x^2}{\sigma^2}} \tag{7}$$

and rewriting the problem as a second order differential equation, yields

$$\frac{d^2\psi}{dx^2} = -\frac{2\mu}{\hbar^2}(E - V(x))\psi \tag{8}$$

Solving the differential equation above with the shooting method were done by connecting the previous (ψ_{i-1}), present (ψ_i), and the next (ψ_{i+1}) wave function (state) following this relation

$$\psi_{i+1} = 2\psi_i - \psi_{i-1} + \Delta_x^{\ 2} f_i \psi_i + O(\Delta_x^{\ 6}) \tag{9}$$

With f_i is the second order differential equation of our interest. By replacing f_i with the right side of Equation (11), yields

$$\psi_{i+1} = 2\psi_i - \psi_{i-1} - \frac{2\mu}{\hbar^2}(E - V(x))\Delta_x^{\ 2}\psi_k \tag{10}$$

The appropriate boundary conditions must be chosen by setting the wave function to be very small in either sides of the potential wall, $\psi(-\infty)=\psi(\infty)\approx 10^{-20}$, and by setting the stopping criterion $\psi_{i+1} - \psi_i = 10^{-20}$ and $E_{i+1} - E_i = 10^{-6}$. The eigenvalues and eigenfunctions of the system then can be calculated numerically as a function of pairing constant.

3. Result and Discusions

The results of numerical method and the perturbative approach (the first and second order correction) on the energy spectrum of QHONGP system are shown in Figure 1, in which the energy is plotted as a function of pairing strength g. In this paper, we assumed that the numerical approach produces a better result (closer to the intended solution) as compared to the perturbative approach which was limited only up to the fourth excited state and the first five terms of the expansion. As the pairing strength grows, the energy of the system in all states is decreased, with the exception at the third excited state (using second order correction approach), indicating a stronger pairing formation. The first order correction of perturbative approach showed a linear decrease of energy, while the solution of the numerical method showed a parabolic-like energy decrease in a similar fashion as reported by others in the quantum harmonic oscillator with a delta-function potential [11,12,16]. Note that by tunning σ, the width constant in Equation (1), the parabolic profile of this energy decrease can be changed, but for simplicity the parameter σ have been set to unity, since in this case, we put more attention to the effect of pairing strength g rather than the potential's width σ.

For the ground state and the first excited state, it was found that the second order correction produces a better result as compared to the first order correction in approaching the result from the numerical method. The different result was obtained in the second excited state, where the first order correction produces better results than the second order correction. The inaccuracy of the second order correction comes from the oscillating sign in the perturbative expansion which depend on the number of terms included in the perturbative expansion. The inaccuracy of the second order correction becomes even larger in the third excited state, as a result of a bigger pairing repulsion in the energy expansion. In the fourth excited state, the inaccuracy between perturbative approach and numerical approach is basically vanished, which might come from the negative energy term larger than the positive term and stabilize the energy fluctuation to bring the energy back to the numerical result. We assume that for the higher level (n>5) the results from both perturbative and numerical approach will be convergent as shown by the result from the fourth excited state. Note that, there is a small bump in the numerical result of the fourth excited state which might come from the numerical artifact.

The energy of the system as calculated by numerical and perturbation method in a small pairing strength region ($g < 4.0$) are almost identical for all states, while for higher pairing strength ($g > 4.0$), the difference between both methods becomes more obvious, especially in the first, second and third excited state. The discrepancy between perturbative and numerical approach at some lower states (n=1,2,3) and high pairing strength likely comes from the limited terms included in the perturbative calculation which resulted in the non-negligible competition between the positive and the negative energy terms in a perturbative approach. Inclusion of more terms in the calculation will probably converge the results of the perturbative approach to the values of the numerical approach.

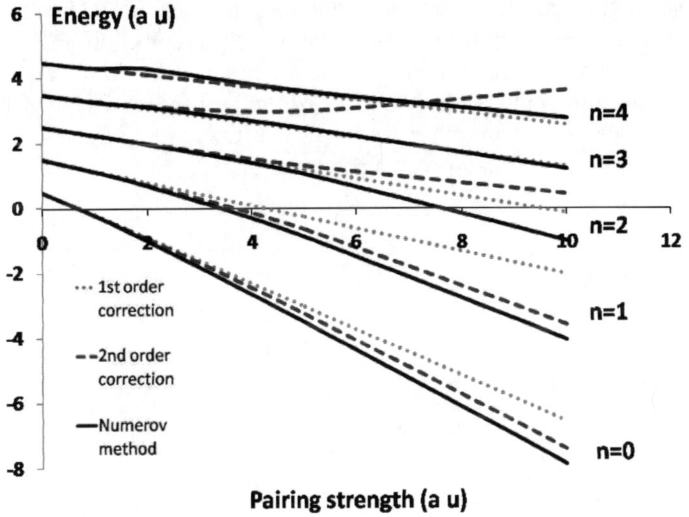

Journal of Physics: Conference Series **1354** (2019) 012019 doi:10.1088/1742-6596/1354/1/012019

Figure 1. The energy of QHONGP system calculated by a numerical (Numerov and shooting) method and perturbative approaches (first and second order corrections) as a function of pairing strength for the ground state and the first four excited states (n=0,1,2,3,4). All units are in arbitrary.

The feature of QHONGP's solution is slightly different than the quantum harmonic oscillator with a delta function potential (QHODP) as discussed by others [11–13] in two ways. Firstly, in QHODP the pairing potential only affecting the even states due to the symmetry of the wavefunction at the equilibrium point, while in QHONGP the pairing potential affecting all (odd and even) states. Secondly, the pairing potential in QHODP reduces the energy of the even states to saturate to the energy of the odd states located at one level below, while in QHONGP no such behavior presents. These differences actually tell us the difference nature of pairing explained by those two models. The QHODP with stronger pairing effect is able to describe the transition from a weak pairing (BCS) to a strong pairing (BEC), while the QHONGP basically only explain the weak pairing formation (BCS or a molecular state) in some lower states due to a shallow pairing potential.

The energetic behavior of the system at extremely large pairing strength approached by the perturbative method even though producing a lower energy (stronger pairing) but it seems irrelevant to discuss since in that situation the perturbing potential is larger than the energy of the unperturbed system and makes the perturbative approach no longer valid.

The effect of the negative Gaussian potential to the wavefunction of the quantum harmonic oscillator was calculated via a perturbative approach for the first order correction only. The probability density ($|\Psi|^2$) of ground state at various pairing strength is shown in Figure 2. As the pairing strength grows, the probability density of the system becomes higher as indicated by the formation of a localized system; a pairing bound state. The probability to find the particles is the highest at the equilibrium position where the two particles were brought closer by Gaussian potential and behave like a single state (a Cooper pair). The use of a stronger pairing strength (g=10, in Figure 2) change the profile of ground state probability density from a single peak into the profile of second excited state with three peaks. This finding is consistent with the occurrence of three negative energies (bound state) in Figure 1 at g=10. It is expected that for extremely large g, more peaks which correspond to the energy less than the ground state energy ($E < 0.5\ \hbar\omega$) will present.

For the excited states, the effect of negative Gaussian potential to the probability density of the system was shown in Figure 3. As the pairing strength gets stronger the probability of particles to be localized at a particular distance is increased, and the stronger pairing strength prefers to enhance the probability density at the distance far from the equilibrium point. An interesting result was found in the third excited state, where the number of the peak which initially four (at g=0) was reduced to two (at g=10.0). In that case, the stronger pairing strength eliminates the localization of particles in the region close to the equilibrium point. Similar features of the probability density (or wavefunction) as shown in Figure 3 were also found by other researchers in the case of Bose–Einstein condensation modeled by a quantum harmonic oscillator with a parabolic trap [17].

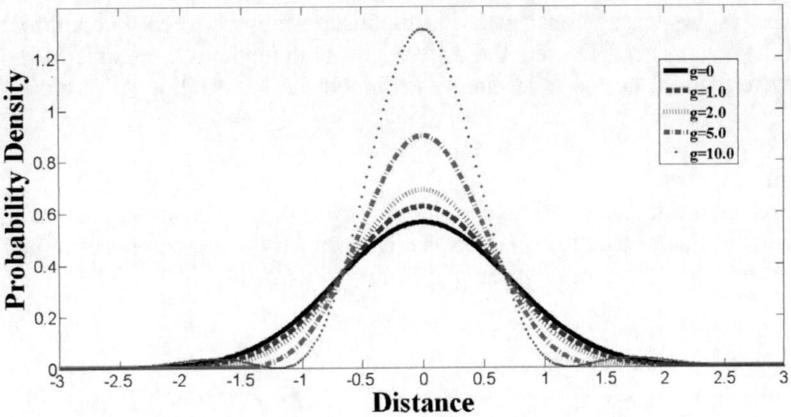

Figure 2. The probability density of ground state as a function of distance for various pairing strength g. All units are in arbitrary.

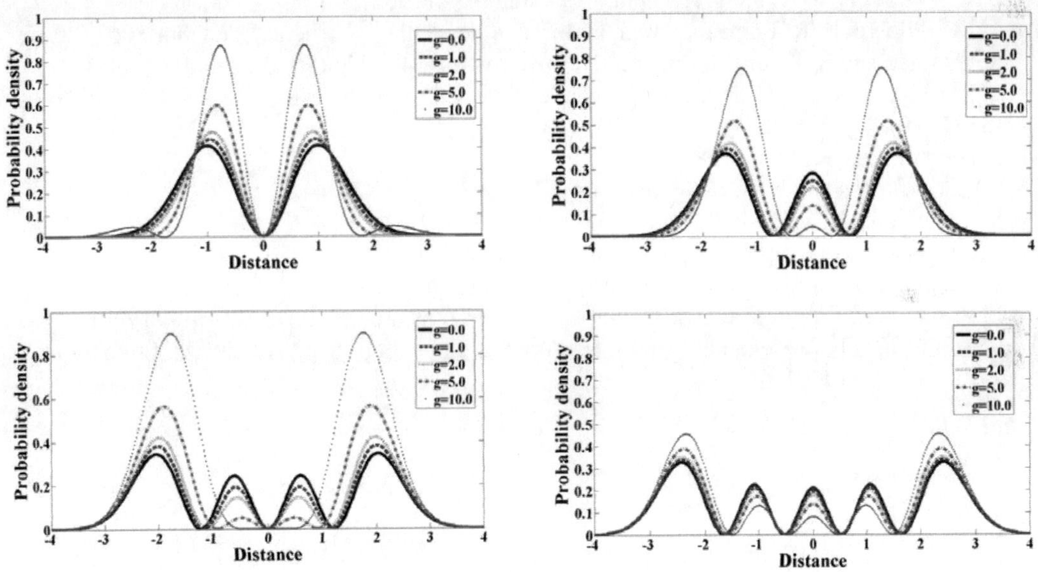

Figure 3 The Probability density of some excited states. (a) first excited, (b) second excited, (c) third excited, (d) fourth excited states. All units are in arbitrary.

4. Conclusion

Throughout this paper, we have demonstrated the ability of a simple quantum harmonic oscillator with negative Gaussian potential in describing the formation of pairing in the spatial basis using numerical and perturbative approaches. The results from the perturbative approaches, as we expected, only make sense in the limit of small pairing strength and in the region close to the equilibrium point. In the higher pairing strength region, the perturbative approach becomes inaccurate in describing the pairing formation since the perturbation energy terms are larger than the unperturbed energy of the system which violate the first principle of perturbation approach (i.e small perturbation around the equilibrium point). In general, the numerical and perturbative approach give similar results in the low pairing constant region, but as the pairing constant grows the discrepancy between both approaches becomes larger and destabilize the system. Despite the differences, both methods were able to show the unique feature of pairing formation, i.e the total energy drop and the localization of particle as the pairing strength grows.

Acknowledgments
Author (TS) expresses gratitude to J. Piekarewicz for fruitful discussions in the early stage of this research. Some parts of this research was funded by The Ministry of Research Technology and Higher Education of The Republic of Indonesia through a Grant No. 1627/IT3.11/PN/2018

References
[1] R. Wallace, Chem. Phys. **11**, 189 (1975).
[2] S. Pramudito, N. W. Sanjaya, and T. Sumaryada, in *AIP Conf. Proc.* (2016).
[3] B. F. Bayman and A. Bohr, Nucl. Phys. **9**, 596 (1958).
[4] A. Sulaiman, F. P. Zen, H. Alatas, and L. T. Handoko, Phys. Rev. E - Stat. Nonlinear, Soft Matter Phys. **81**, (2010).
[5] T. Gao and Y. Chen, Phys. A Stat. Mech. Its Appl. **468**, 307 (2016).
[6] P. M. Stevenson, Phys. Rev. D **30**, 1712 (1984).
[7] J. D. Swalen and J. A. Ibers, J. Chem. Phys. **36**, 1914 (1962).
[8] A. M. Halpern, B. R. Ramachandran, and E. D. Glendening, J. Chem. Educ. **84**, 1067 (2007).
[9] A. Gharaati and R. Khordad, Superlattices Microstruct. **48**, 276 (2010).
[10] B. Szafran, J. Adamowski, and S. Bednarek, Phys. E Low-Dimensional Syst. Nanostructures **5**, 185 (1999).
[11] S. H. Patil, Eur. J. Phys. **27**, 899 (2006).
[12] J. Viana-Gomes and N. M. R. Peres, Eur. J. Phys. **32**, 1377 (2011).
[13] T. Sumaryada, B. Maha Putra, and S. Pramudito, Eur. J. Phys. **38**, (2017).
[14] J. J. Sakurai and J. Napolitano, *Modern Quantum Mechanics* (2010).
[15] V. Fack and G. Vanden Berghe, J. Phys. A. Math. Gen. **20**, 4153 (1999).
[16] T. Busch, B.-G. Englert, K. Rzażewski, and M. Wilkens, Found. Phys. **28**, 549 (1998).
[17] Y. S. Kivshar, T. J. Alexander, and S. K. Turitsyn, Phys. Lett. Sect. A Gen. At. Solid State Phys. **278**, 225 (2001).

Three Dimensional Numerical Modeling of Current and Temperature Distribution in the Estuary of Palu River

Mohammad Lutfi

Department of Petroleum Engineering, STT MIGAS, Balikpapan, 76127, Indonesia
E-mail: Lutfi_plhld@yahoo.co.id

Abstract. Three dimensional numerical model of current and temperature distribution in the estuary of Palu river was discussed in this article. The simulation was conducted using estuary coastal ocean model and sediment transport (ECOMSED) and geographical information system (GIS). The first simulation was conducted by assuming no discharge and the second simulation using combination of tidal and river discharge. The first simulation runs smoothly, where the results revealed that the tidal current velocity on the horizontal and vertical cross sections for one tidal cycle ranged from 0 m/s to 0.03 m/s. Whereas a run time error occurred during the second simulation due to the calculation of nonliniear advection term parameter in momentum equation. The error problem could be solved by imposing a low river discharge data into the model. The visual appearance of current circulation at the river mouth, calculated surface elevation, and temperature distribution on the horizontal and vertical cross sections revealed a good agreement with the primary and secondary data.

1. Introduction

The estuary of Palu river is a place for fishermen to earn a living. Industrial developments along the river banks have triggered waste disposal activities into the aquatic environment. The problem that frequently occurs is that during the rainy season water in the river overflows, it has led the pollutant enters the houses of some residents near such area. Therefore, an integrated environmental management is needed. An approach to solve this environmental problem is by collecting information that can show the pattern of pollutants dispersion through the current circulation prediction in the estuary. This information can be obtained through continuous monitoring activities. However, this activity requires a high cost, therefore, numerical model is one of solutions that can be used to solve this problem.

A predicting of hydrodynamics in coastal zone due to hydrometeorological activity using numerical model is also important for monitoring water quality and early warning system related to the water crisis in Palu city [1, 2]. Therefore, current circulation and temperature distribution model is needed to be made. One of the three-dimensional numerical models that can be applied to analyze the dynamics of the estuary physical oceanography in the region is the estuary coastal ocean model and sediment transport (ECOMSED) [3]. This model has been used by several researchers to simulate hydrometeorology phenomena in the estuary [4], tide current in the bay [5], a typhoon storm surge [6], inundation [7], three dimensional hydrodynamic over bed forms in open channels [8], etc. Where the

Content from this work may be used under the terms of the Creative Commons Attribution 3.0 licence. Any further distribution of this work must maintain attribution to the author(s) and the title of the work, journal citation and DOI.
Published under licence by IOP Publishing Ltd

simulation results revealed that there is no significant difference with the observed data and theoretical model.

Numerical calculation in the estuary of Palu river became unstable due to the water depth rapidly change from nearshore to the open sea based on the previous researches [9, 10], where bathymetry data obtained based on the contours analysis of the sea map of the Palu bay and general bathymetric chart of the oceans (GEBCO) digital atlas [11]. Modification of bathymetry data will change the current velocity on the steep area and vicinity, therefore, in this article, the simulations were conducted without changing bathymetry data which is obtained from interpolation.

This research aims to make the current circulation and temperature distribution model in the estuary of Palu river. The results are expected to provide information to anticipate the pollution, so that losses can be prevented before a disaster occurs.

2. Methods

Three dimensional model for current and temperature distribution which is generated by tide and river discharge are simulated separately to reveal the dynamics of current and temperature in the estuary of Palu river.

Topography map, IKONOS image, and google earth image are used as reference points for geometric correction in simulation area. The more reference points, the error during geometric correction can be reduced, so that the model domain has a high level of precision and accuracy.

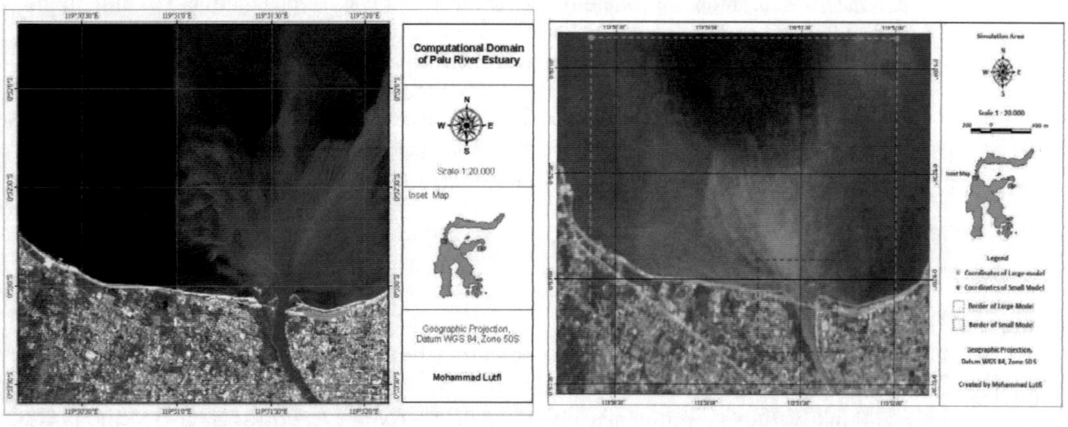

(a) **(b)**

Figure 1. Map of simulation area based on the satellite image of (a) IKONOS and (b) google earth

The simulation area is divided into 125 x 125 grids horizontally and 11 layers vertically with the different thickness. The parameter settings are as follows: $\Delta x = 25\,\text{m}$, $\Delta y = 25\,\text{m}$, $\Delta t = 1\,\text{s}$, the simulation was running 360 hours continuously, constant temperature value of 28^0 at the upstream (river) and 29^0 at the downstream (sea), constant salinity value of 0 psu at the upstream and 34 psu at the downstream, river depth obtained from the Department of Public Works of Palu City in year 2008, bathymetry data obtained from interpolation of the sea map of Palu bay, and six tidal harmonic constants obtained from hydro oceanographic office (DISHIDROS) [12].

The first simulation was conducted by assuming no discharge to reveal the current velocity due to the tidal effect. Second simulation were conducted for the river discharge of 36 m^3/s and 2 m^3/s. in order to see clearly the current and temperature distributions at the delta and vicinity, the simulation results were displayed in high resolution.

Verification of model was conducted by comparing the simulation results to the secondary data from DISHIDROS) for water elevation and field observation of temperature distribution was carried out at the same locations with simulation area on the vertical cross section.

3. Results and Discussions

3.1. Depth interpolation

The interpolation was conducted using Arcgis and Tecplot based on the sea map of the Palu bay. The water depth of the Palu bay shown in Fig. 2 was obtained by digitizing the map and then creating interpolation using the kriging interpolation method. The result shows that the river depth ranging from 1.5 m to 3 m and the bathymetry ranging from 0 m to 167.83 m, where the depth rapidly change in the eastern part of the sea

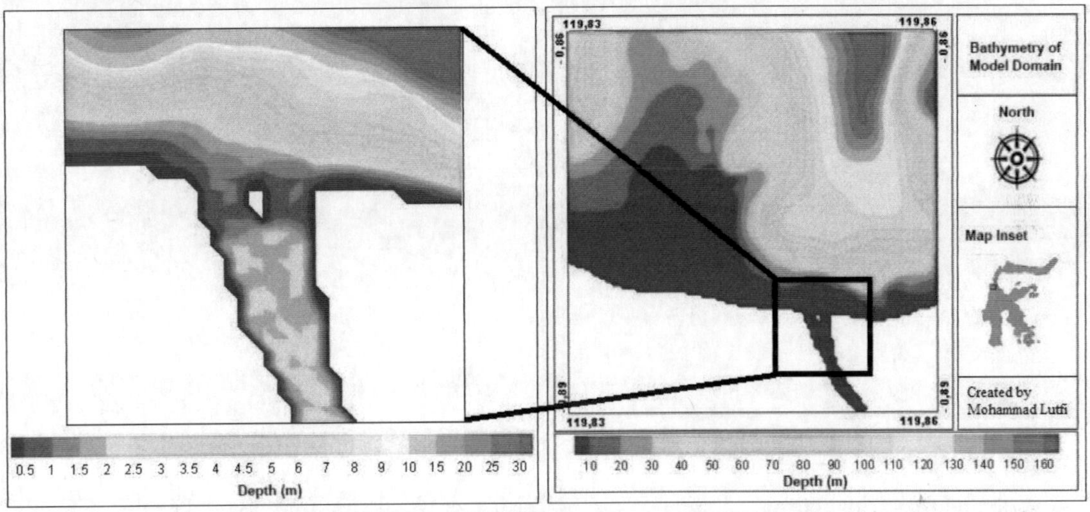

Figure 2. Depth interpolation of simulation area

3.2. Run time error

Run time error occurred during second simulation with discharge of 36 m^3/s due to the calculation of nonliniear advection term parameter in momentum equation. The problem could be solved by turning off convective acceleration. It means that the analytical solution can be obtained by solving the differential equation using techniques of integration for elementary functions [13, 14]. However, the simulation results will be unrealistic to describe the nonlinear phenomena, therefore, the error problem was solved by imposing the river discharge of 2 m^3/s as input data into ECOMSED. The instability of numerical calculation due to the river discharge of 36 m/s^3 could be solved by reducing the gradient of the depth slope of bathymetry on steep areas [9]. However, the current velocity will change because of the velocity dependent on the rate of depth change.

Many numerical calculations have generally the computational stability condition. In the ECOMSED, the Courant-Friedrichs-Levy (CFL) stability condition should be satisfied as one of stability criterion. According to the CFL condition, numerical stability is dependent on water depth and maximum flow velocity. Therefore, numerical calculation become unstable when water discharge is larger. Other, numerical instability is dependent on the rate of depth change and velocity change. When maximum depth is 168.33, CFL conditions are internal mode time step (DTI) < 8.9s and external mode time step (DTE) < 0.22s.

3.3. Tidal current and surface current

Tidal current simulation was conducted for the first simulation by assuming no discharge to test the stability of ECOMSED and to reveal the current velocity that is only generated by tidal, so that the effect of tidal and river discharge at the estuary can be revealed. The simulation was carried out by imposing the tidal harmonic constants as a generated force on the open sea boundary was extracted from harmonic analysis, where the simulation runs smoothly without any error.

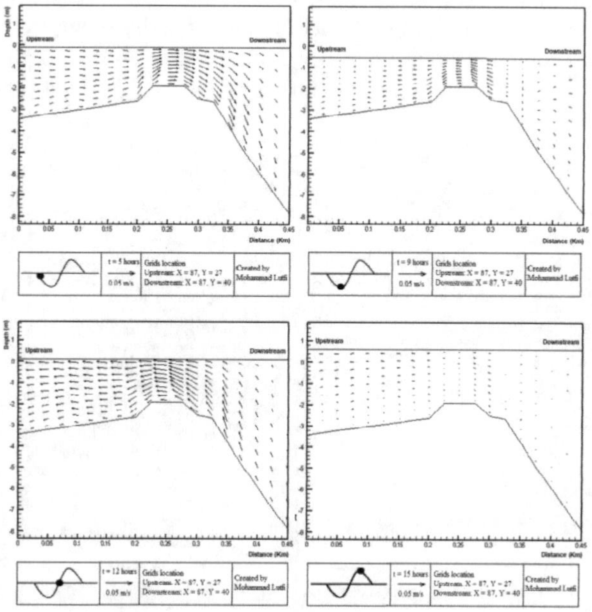

Figure 3. Tidal current simulation on the vertical cross section for one tidal cycle

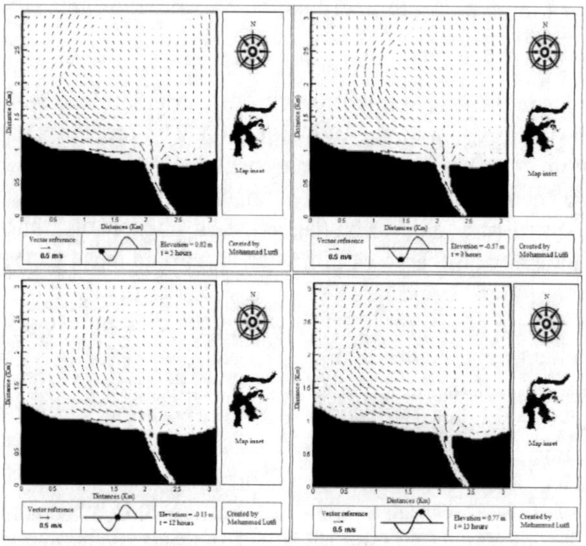

Figure 4. Surface current simulation for one tidal cycle

Journal of Physics: Conference Series **1354** (2019) 012020 doi:10.1088/1742-6596/1354/1/012020

Fig. 3 shows that the current pattern reverses during low slack water and high slack water conditions, it flows alternately in approximately opposite directions. At low slack water, the current pattern moves from downstream to upstream, whereas at high slack water flows from upstream to downstream. The current velocity in this conditions varies from 0 m/s to 0.03 m/s. The current velocity at ebb condition begins to slow down and reach the minimum velocity at the flood condition with the velocity varies from 0 m/s to 0.008 m/s.

The direction of current will change dynamically depend on the position and geometry of delta at the river mouth. The visual appearance of Fig. 4 shows the direction of current movement based on the simulation results has similar patterns with the google earth image in Fig 1(b), where the surface current velocity in the estuary varies from 0 m/s to 0.54 m/s.

3.4. Temperature distribution

Fig. 5 depicts the high resolution of simulation area around the delta and vicinity. The surface temperature distribution on the horizontal cross section from the upstream to downstream varies from 28^{0}C to 29^{0}C. This is caused by a constant input data of surface temperature in ECOMSED, where in upstream and downstream are 28^{0}C and 29^{0}C respectively. It can be seen in Fig. 5, that surface temperature distribution follows the patterns of surface current from the river.

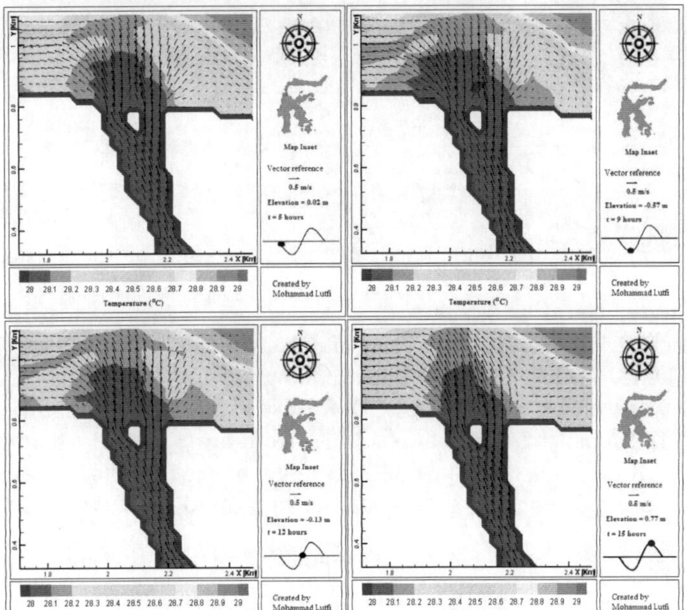

Figure 5. Temperature distribution on the horizontal cross section for one tidal cycle.

3.5. Verification of temperature distribution

Fig. 6 depicts the temperature distribution on the vertical cross section from the river to the sea. The boundary between the fresh water from the river and saline water from the sea can be seen clearly, where the fresh water that has higher temperature lies above the saline water that has lower temperature due to the different salinity between fresh water and saline water. The lower salinity of the fresh water corresponds to the lower density, so that it lies above the layer of saline water from the sea that has higher density. These conditions in line with a theoretical model, where the water temperature decreases with the depth. At ebb and flood conditions, the difference of water temperature between the near-surface and near-bottom rarely exceeded 1°C [15].

The observation was conducted on three observation stations on the vertical cross section from the bottom to the surface around the delta, water temperature at low slack water, ebb, and high slack water condition was 28^{0}C from bottom to surface. It reveals a uniform condition from the lowest layer to the surface. At flood condition, water temperature was 29^{0}C at the lowest layer. This condition similar to the simulation results in Fig. 6, where the temperature distribution varies from 28^{0}C at the surface and 28.8^{0}C at the bottom.

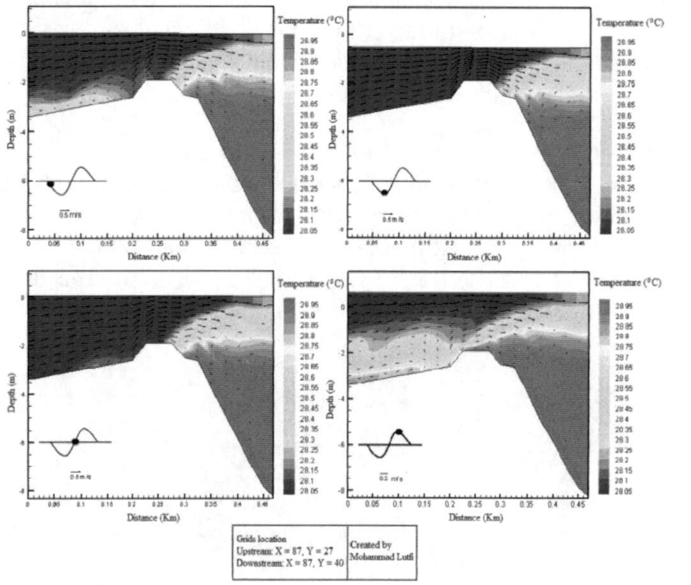

Figure 6. Temperature distribution on the vertical cross section for one tidal cycle.

3.6. Verification of water elevation

Verification was conducted by comparing secondary data on the tidal area of the Palu bay from DISHIDROS and ECOMSED. The simulation was running for 360 hours continuously. According to Fig. 7, comparison of water elevation between ECOMSED and DISHIDROS shows a good agreement, where there exist a similar patterns for both results. According to the tidal harmonic constants, the Formzahl value is 0.29. It means that the type of tide in Palu bay included of the mixed tide prevailing semidiurnal, where on one day, ebb and flood conditions occurred twice with the different amplitude and period.

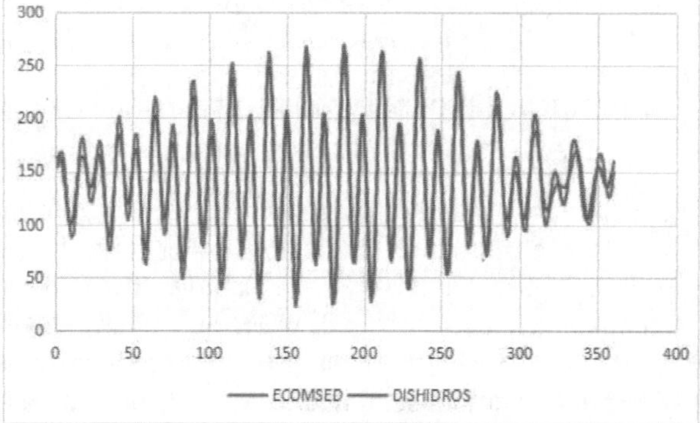

Figure 7. Comparison of water elevation between ECOMSED and DISHIDROS

4. Conclusions

The first simulation revealed that the tidal current velocity on the vertical cross sections varies from 0 m/s to 0.03 m/s. Run time error occurred during second simulation due to nonliniear advection term parameter in momentum equation, the problem could be solved by imposing a low river discharge into the ECOMSED. The visual appearance of current circulation at the river mouth and vicinity, calculated surface elevation, and temperature distribution on the horizontal and vertical cross sections revealed a good agreement with the image satellite obtained from google earth, temperature observation, and water elevation obtained from DISHIDROS.

The results are expected to predict the spread of pollutants and sea surface temperature with a low river discharge during ENSO. This model can be used for a large river discharge by reducing the gradient of the depth slope of bathymetry on steep area and increasing the resolution of space and time near such area. The next research topic will be conducted using nesting model for a better resolution. The current patterns and temperature distribution are strongly influenced by several parameters such as discharge and water elevation. Nevertheless, a comprehensive simulation need to be conducted by considering the effects of wind and wave. In order to see clearly the dynamics of the shoreline moving over tidal flat, then the wet and drying scheme (moving boundary) need to be applied into the numerical model.

Acknowledgement

The author would like to thanks the International Development and Cooperation (IDEC), Hiroshima University for the environment simulator laboratory facilities during this research. Special thanks also goes to Dr. Mishima and Drs. Mr. Mardeli Jandja for the helpful discussions and encouragement.

References

[1] Sulaiman B, Bambang A N, Purnaweni H, & Lutfi M. 2018. The effect of mangrove on fish catch using *belat* at Teluk Pemedas and Sanipah of Kutai Kartanegara regency, east Kalimantan province, Indonesia. *E3S Web of Conferences*. **31** 08029.

[2] Tyree J, Ven Johnson S, Pascua M C, Rahaman M M, Tenorio A L, & Lutfi M. 2013. Effectiveness of camanava flood control project: A case study of selected flood control structures during typhoon gener and monsoon rains in August 2012. J. of IDEC, **19**, 3 pp. 131-45. Doi: http://doi.org/10.15027/35045.

[3] HydroQual, Inc. *A Primer for ECOMSED*, users manual, Mahwah, N. J. 07430, USA, 2002.

[4] Kuang C P, Sun B, & Liu S G. 2009. Approach of moving boundary and its application in 3D tidal current simulation of Changjiang River estuary based on ECOMSED model. *Acta Oceanol. Sin*, **28** pp 78-85.

[5] Du P J, Hu K L, Kong Y Z, & Ding P X. 2007. Application of ECOMSED model to the simulation of Hangzhou Bay tide current. *Acta Oceanol. Sin*, **29** 1 pp 7-16.

[6] Nie H, Sun Z, & Xie C. 2012. Simulating a typhoon storm surge using a nested Ecomsed model. *Procedia Engineering*, **31** pp 775-780.

[7] Zhao H L & Zhang Y F. 2010. Development of inundation scheme for ECOMSED. *Journal of Waterway and Harbor*, **5**, 029.

[8] El Kheiashy K, McCorquodale J, Georgiou I, & Meselhe E. 2010. Three dimensional hydrodynamic modeling over bed forms in open channels. *International Journal of Sediment Research*, **25** 4 pp 431-440.

[9] Lutfi M, Afifah R S, Sulaiman B, and Risna. 2017. *Indian J. of Sci. and Tech,* **10** 29. Doi: 10.17485/ijst/2017/v10i29/116502.

[10] Lutfi M. 2015. Three dimensional model for cohesive sediment transport in estuary of Palu river. *Journal of Mitra Sains*. **3** 1 pp. 78-87.

[11] IOC, IHO & BODC. Centenary Edition of the GEBCO Digital Atlas, published on CD-ROM on behalf of the Intergovernmental Oceanographic Commission and the International Hydrographic Organization as part of the General Bathymetric Chart of the Oceans, British Oceanographic Data Centre, Liverpool, U.K, 2003.

[12] Indonesian National Armed Forces. 2005. *The Book of Tidal in Indonesia Archipelago*. Hydro Oceanographic Office, Jakarta (in bahasa Indonesia).

[13] Jandja M & Lutfi M. 2018. The Five columns rule in solving definite integration by parts through transformation of integral limits. *J. Phys: Conf. Ser.* **1028** 012109. IOP Publishing. Doi:10.1088/1742-6596/1028/1/012109

[14] Lutfi M. 2018. The effect of gravitational field on *brachistochrone* problem. *J. Phys: Conf. Ser.* **1028** 012060. IOP Publishing. Doi:10.1088/1742-6596/1028/1/012060

[15] Geawhari M A, Huff L, Mhammdi N, Trakadas A, & Ammar A. 2014. Spatial-temporal distribution of salinity and temperature in the Oued Loukkos estuary, Morocco: using vertical salinity gradient for estuary classification. *SpringerPlus*, **3** 1 p. 643.

Finite Difference Method For Laplace Equation In Irregular Domain

E. Juarlin

Department of Physics, Faculty Mathematics and Natural Science, Hasanuddin University. Makassar. South Sulawesi. Indonesia.

E-mail: ekojuarlin@unhas.ac.id

Abstract. Laplace equation on irregular domain has been solved by finite difference method. The program for calculating distribution of electric potential inside has been developed. Symmetrical Dirichlet boundary condition and Cartesius Coordinates are applied. A trapezoid and a quarter of a circle are chosen as irregular domain example. The contour of electric potential shows symmetrical result to its domain. The convergence and stability output program shows a good result.

1. Introduction

Electric potential is a scalar variable, appearing from one or many electric sources that put in typical media. Electric potential is special property of electric field to reduce a vector problem down to a scalar problem. Electric potential is an useful physical quantity in many fields concerned with electromagnetism.

Laplace equation is a second order partial differential equation (PDE) that appears in many areas of science. In physics it uses to find some physical quantities, like temperature, electrics potential, amplitude of wave and so on Solution of this equation, in a domain, requires the specification of certain conditions that the unknown function must satisfy at the boundary of the domain. The function itself is specified on a part of the boundary called Dirichlet boundary but the normal derivative of the function is specified on a part of the boundary, called Neumann boundary condition.

The easiest method of numerical techniques to get solution of Laplace equation is Finite Difference Method (FDM). There are commonly two reasons why FDM is used. First, the shape is irregular. Second, the solution can not be calculated analytically. FDM proceeds by replacing the derivatives in the differential equations with finite difference approximations according to Taylor expansions [1]. A mathematical relation between neighbourhood points inside is calculated. The solution can be calculated simultaneously or iterative. Widely, FDM has been implemented in many field, like vibration [2], thermal spreading of geothermal systems [3], electronic field [4].

2. Laplace Equation in FDM

The Laplace equation for electric potential is

$$\frac{\partial^2 V}{\partial x^2} + \frac{\partial^2 V}{\partial x^2} = 0 \tag{1}$$

Content from this work may be used under the terms of the Creative Commons Attribution 3.0 licence. Any further distribution of this work must maintain attribution to the author(s) and the title of the work, journal citation and DOI.
Published under licence by IOP Publishing Ltd

The second derivatives for two dimensions can be approximated as

$$\frac{\partial^2 V}{\partial x^2} \approx \frac{V(i-1,j) - 2V(i,j) + V(i+1,j)}{\Delta x^2} \tag{2}$$

So, two dimension FDM approximation for Laplace equation is

$$\frac{V(i-1,j) - 2V(i,j) + V(i+1,j)}{\Delta x^2} + \frac{V(i,j-1) - 2V(i,j) + V(i,j+1)}{\Delta y^2} = 0 \tag{3}$$

From equation 3, after all points are made into, finally a matrix containing relation electric potential between points and a column matrix containing boundary value are constructed. The solution for a set of linear algebraic equations is calculated by Gauss Seidel iteration until electric potential value is convergent.

3. The Algorithm for Numbering Points

Numbering points inside shape needs a special technique because shape is irregular. The numbering technique for both domains is almost same.

A bigger rectangle than domain is constructed outside trapezoid. Then, trapezoid is constructed by four lines mathematically determined by four equations of all circumference line. One circumference line is negative gradient, one circumference line is positive gradient and two others are zero gradient. Points are discretized horizontally and vertically along to Δx and Δy. There are points outside trapezoid. Perhaps, there is one or more points nearest circumference and there is one or more points lying exactly on circumference. The domain is cut so its shape is formed from a collection of horizontal and vertical grid lines having same interval distance. On every horizontal grid line, there are two end points inside trapezoid such as one point is near to negative gradient line and one point is near to positive gradient line. On every vertical grid line, there are two end points inside trapezoid such as one point lies on x-coordinate and one point lies on top of trapezoid. The rest points are calculated points. All points have been classified.

In quarter circle, the technique for numbering is similar to trapezoid numbering. Quarter circle lies on first quadrant. There are three circumference lines that one vertical line lying on y coordinate, one horizontal line lying x coordinate and one as an arc. Vertical and horizontal grid lines are made up into domain such as every intersection has a discrete point. After putting discrete points inside, there must be an end point on top at a certain x and an end point on right at a certain y. On all end points, Dirichlet boundary condition is applied. The rest points are calculated points.

4. Simulation Results

Domain is figured in figure 1.

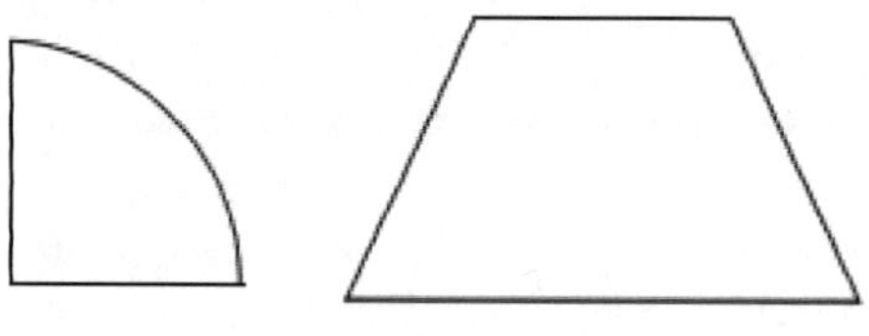

Figure 1. Domain of a square of circle (left) and a trapezoid (right)

The center of coordinate is as point of center of a circle in this paper. Electric potential source 10 V is placed on circumference of quarter circle line. Two functions are used such as

$f(x, y) = 10$ on first quadrant points which obeys equation $x^2 + y^2 = r^2$ and $f(x, y) = 0$ for $x = 0$ and $y = 0$ which are applied as boundary conditions. Radius of a square of a circle is 10, and value of X interval and Y interval is 1. The contour of electric potential for a quarter circle is shown on figure 2.

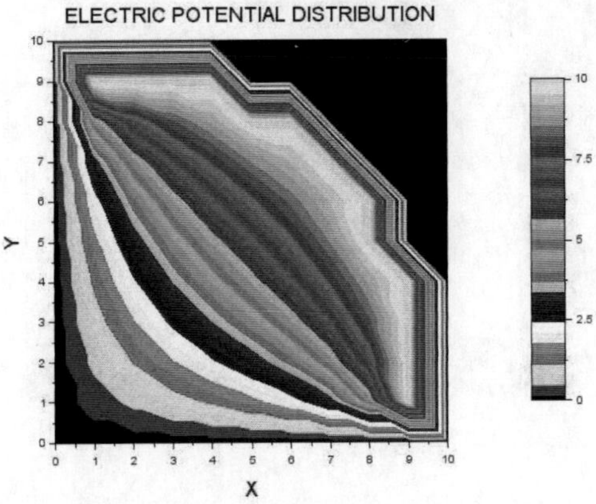

Figure 2. Electric Potential Distribution in Square of Circle with Radius 10

Figure 2 displays the electric potential distribution is symmetry to line y = x. Electric potential becomes greater as it becomes near to arc. The domain shape is not a perfect quarter circle because its edge cutting effect. The electric scalar potential in all points reaches convergence in about 100 iterations with 56 calculated points and simulation results a good stability as in figure 3.

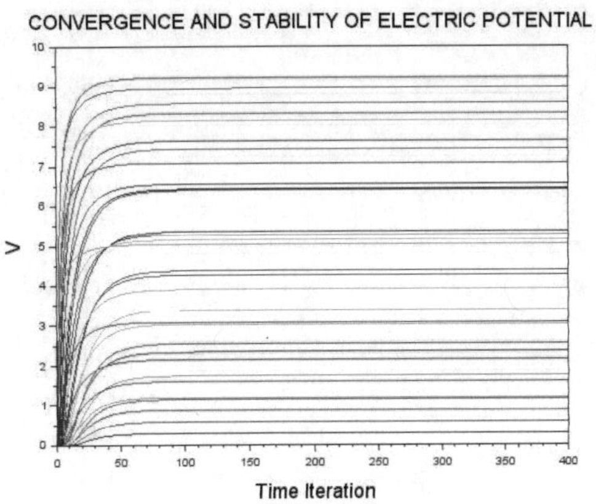

Figure 3. Convergence and Stability of Square of Circle with Radius 10

In a trapezoid, interval one segment for both axes is equal which is 0.5, 1 and 2. Length on bottom is 10 and length on its top is 6 and length on its height is 6. Base of trapezoid lie on x axis. The isosceles trapezoid is taken as example. The value of electricity potential is 10 at bottom and base but 0 at hypotenuse. The electric potential distribution at interval 0.25 is illustrated in figure 4.

Journal of Physics: Conference Series **1354** (2019) 012021 doi:10.1088/1742-6596/1354/1/012021

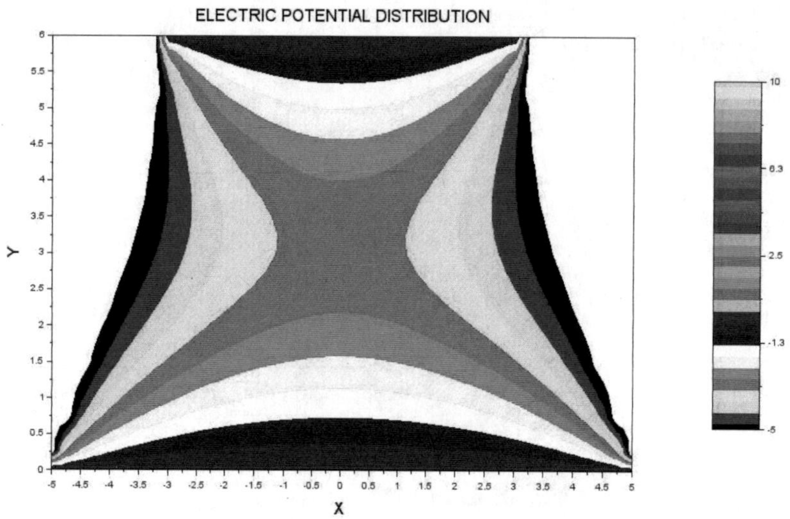

Figure 4. Electric Potential Distribution in Trapezoid with Interval 0.25

As a symmetrical boundary condition to y axis, the simulation shows symmetrical result. From center of trapezoid, electricity potential becomes greater to the base and top but becomes lower to the hypotenuse. This case has 39 calculated points and reaches stability after about 250 iteration. The convergence and stability show good result as showed in figure 5.

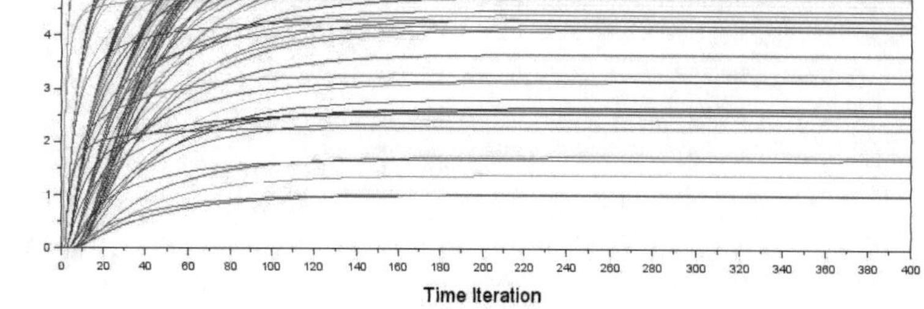

Figure 5. Convergence and Stability of Trapezoid with interval 0.25

5. Conclusion

We have constructed a simulation of electricity potential on irregular domain with symmetrical Dirichlet boundary condition and symmetrical domain. The simulation result is agreed with theorem. Convergence and stability as analyzed variable to determine quality of simulation shows good result.

Acknowledgments

First I would like to express my special thanks to IOP for its oppurtunity to publish my research. Second I would say thank you very much to CTPNP 2018 reviewer and editor on their work of review my paper so a good paper is accepted in IOP journal.

References

[1] Randall J. LeVeque 2006 *Finite Difference Methods for Differential Equations* (University of Washington)
[2] Scolastika Lintang Renganis Radityani, Sudi Mungkasi 2017 *Journal of Physics: Conf. Series* **909** 012044
[3] Joko Sampurno, Apriansyah, Riza Adriat, Irfana Diah Faryuni, Wahyu Srigutomo 2017 *Journal of Physics: Conf. Series* **877** 012012
[4] Gefei Yu, Xuewen Li 2013 *Journal of Physics: Conference Series* **418** 012016

AUTHOR INDEX